OTHER BOOKS BY JOHN BRIGGS

The Fire in the Crucible

The Alchemy of Creative Genius, by John Briggs

Turbulent Mirror

An Illustrated Guide to Chaos Theory and the Science of Wholeness,

by John Briggs and F. David Peat

Looking Glass Universe

The Emerging Science of Wholeness,

by John Briggs and F. David Peat

Metaphor

The Logic of Poetry, by John Briggs and Richard Monaco

A NEW AESTHETIC OF ART, SCIENCE, AND NATURE

JOHN BRIGGS

A TOUCHSTONE BOOK PUBLISHED BY SIMON & SCHUSTER
NEW YORK LONDON TORONTO SYDNEY TOKYO SINGAPORE

F R A C T A L S

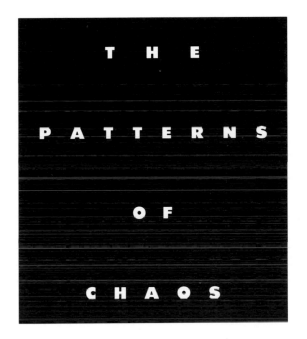

THE
PATTERNS
OF
CHAOS

TOUCHSTONE

Simon & Schuster Building
Rockefeller Center
1230 Avenue of the Americas
New York, New York 10020

Designed by Bonni Leon

Manufactured in the United States of America

10   9   8   7   6   5   4   3   2   1

Library of Congress Cataloging-in-Publication Data
Briggs, John.
    Fractals : the patterns of chaos / John Briggs.
        p.   cm.
    Includes bibliographical references and index.
    1. Fractals.   I. Title.
QA614.86B75   1992
514′.74—dc20   92-16886
                                                              CIP

0-671-74217-5

Grateful acknowledgment is made to the following for
permission to reprint previously published material:

Alfred Dorn. "Snowflake," *The Diamond Anthology.*
Cranberry, N.J.: A. S. Barnes and Company for The Poetry Society of America, 1971.
Edward Berko. *On the Nature of Fractalization.* copyright © 1991 Edward Berko. Publisher: *Ligeia Art Journal, Fractalisms* issue 1991. Ed. Giovanni Lista, Paris.

TO

JEFF,

RICKI,

AND

DEVIN

(MAY HE JOYOUSLY

INHERIT THIS FRACTAL WORLD)

A C K N O W L E D G M E N T S

My greatest gratitude for help on this book must go to its many illustrators who are listed at the back. They have been unfailing in providing me with information, marvelous pictures, and patience. In locating fractal artists, several people aided me: Denis Arvay of IBM in Yorktown Heights, New York; Cliff Pickover also of IBM in Yorktown, a great networker as well as a great fractalier; Professor Milton Van Dyke, Department of Mechanical Engineering, Stanford University; fractalist painter Carlos Ginzburg, another great networker. I want to acknowledge the enthusiastic assistance of art critic Klaus Ottmann; Mark Eustis of the Earth Observation Satellite Company in Lanham, Maryland; and Douglas Smith, curator of the Boston Museum of Science, who, among other advice, taught us how to photograph video chaos. I thank my colleagues at Western Connecticut State University, Professors Hugh McCarney (who actually filmed the video chaos), Margaret Grimes, Bill Quinell, Kalpataru Kanungo, and Susan Maskel for their interest and help on the project.

I thank Karen Holden of Simon & Schuster for initiating the idea of a book on fractals, and Heidi von Schreiner at Touchstone for taking it over and seeing it through with such professionalism, grace, and forbearance (thanks, Heidi). Bonni Leon for her wonderful design of this book, given all the strange constraints I placed on her, and for bringing both it and her new son to term at the same time. I thank especially my assistant Kristina Masten, without whose hours of research this book would simply not have been possible, and Carol Zahn, whose early research helped to locate many of my sources of fractals. I thank David Peat for his kindness in reading the manuscript and relieve him entirely from any responsibility for errors therein. I thank my wife, Joanna, as always, for putting up with my moods and long hours while writing. Last but not least, I thank my agent Adele Leone for her dogged determination and nurturing to get the project done. And Benoit Mandelbrot for his early advice and invention of fractals.

C O N T E N T S

INTRODUCTION 13

A PLANET OF LIVING FRACTALS 35

OF CAMELS, STRAWS, AND FRACTALS 43

THE FRACTALS AND CHAOS OF OUTER SPACE 49

OUR WEATHER TODAY IS CHAOS 55

BETWEEN THINGS: FRACTAL DIMENSIONS 61

THE HAUNTING MANDELBROT SET 73

FRACTAL MATH IMITATIONS, BOTH FANCIFUL AND REAL 83

CHAOS AND SYMMETRY HYBRIDS 93

CHAOS SCULPTS FRACTAL LANDSCAPES 99

SPIRALS, SOLITONS, AND SELF-ORGANIZING CHAOS 107

FEEDBACK AND ITERATION: THE HEARTBEAT OF CHAOS 115

THE HUMAN BODY IS A FRACTAL CREATION 123

THE FOLDED ORDER OF TURBULENCE 131

VISUALIZING CHAOS AS A STRANGE ATTRACTOR 137

THE ART OF ABSTRACT IMAGES FROM FRACTAL MATH 147

THE NEW GEOMETRY OF IRREGULARITY 157

GREAT ART'S FRACTAL SECRETS 165

CODA: LIVING WITH UNPREDICTABILITY'S SHAPES 179

CREATING FRACTALS ON HOME COMPUTERS 182

CONTRIBUTORS' BIOGRAPHIES 184

SUGGESTED READING 187

IMAGE CREDITS 188

INDEX 190

We are in the beginning of a major revolution. . . . The whole way we see nature will be changed.
—Joseph Ford, physicist, Georgia Tech University.

"The forecast," said Mr. Oliver, turning the pages till he found it, "says: Variable winds; fair average temperature; rain at times." . . . There was a fecklessness, a lack of symmetry and order in the clouds, as they thinned and thickened. Was it their own law, or no law, they obeyed?
—Virginia Woolf, *Between the Acts.*

Chaos and fractals are nonlinear phenomena, so you are hereby invited to avoid reading this book linearly. Try weaving your own fractal path through the text. Perhaps you started to do that when you first picked the book up. Jumping around might seem a little chaotic, but that's the pattern under discussion here. The button icons you'll find at the end of each chapter are suggestions about where you might jump next in order to learn more about some fractal/chaos topics closely related to the material you just read. For example, this is the button for the Mandelbrot Module. The other buttons are listed below.

· BIOFRACTALS · NONLINEARITY · SPACE · WEATHER · FRACTAL DIMENSIONS · MANDELBROT SET · IMITATIONS · HYBRIDS · FRACTAL LANDSCAPES · SELF-ORGANIZATION · FEEDBACK · FRACTAL BODY · TURBULENCE · VISUALIZING CHAOS · MATH ART · NEW GEOMETRY · ART SECRETS

•

We tend to think science has

explained everything when it

has explained how the moon

goes around the earth. But

this idea of a clocklike

universe has nothing to do

with the real world.

—JIM YORKE, University

of Maryland physicist

who coined the term

"chaos."

•

INTRODUCTION

Obvious and Hidden Order: Chaos, Fractals, and a New Aesthetic

Everybody talks about the weather; it's one thing we have in common. On a given afternoon, sunshine may fall on our porch while a resident in another part of town may have falling rain. But weather is the phenomenon we share. With its variability, general dependability, and moment to moment unpredictability, weather infiltrates our schedules, sets or undermines our plans, affects our moods, and unites us with the environment and each other. Weather is also an example of a mysterious order in chaos.

Some other examples: The pattern created by boulders tumbled over in a glacial landscape, poking from the soil, spotted with lichens and moss. Trees sprouting out of a glade, random branches and twigs

These haunting self-similar forms fell in the backyard of photographer Joseph Cantrell. Fractals record what happens in the transition zones between order and chaos. The leaves of this random bouquet lie in the zone between life and death. Cantrell's lens reveals the aesthetic order in the haphazard grouping and unites the viewer with these fractal objects.

The "fallen leaves" from a collapsing algal cell were caught by biologist Peter Siver. Though they don't appear translucent here, these surreal plates are actually glass scales which the microscopic freshwater algae make out of sand, and secrete and attach in spirals all over their body in order to protect them and let sunlight through to their chlorophyll. After collecting the algae, Siver dried

them on a piece of tinfoil. The glazing of these algae collapsed in a random pattern. The arrangement, which is neither Euclidian nor symmetrical, is irregular and fractal. Siver calls this particular composition "Barn Owls" and dubs his silicone-coated algae "aquatic snowflakes." Real snowflakes, it turns out, are also fractal.

The minuscule plates of the algae also illustrate the fractal scaling properties of nature. In a tiny pond drop swims a world within our world. In the detail of that world is yet another. Our own bodies are collections of worlds within worlds at finer and finer scales. This is a key fractal idea.

tangled together. Swallow scattering into a field like a handful of thrown dust, rising in a riotous twittering, then gathering and flying off in an organized flock. A lightning bolt fracturing the sky.

Most people find the haphazard profusions of nature so intensely pleasing, even spiritually profound, that it seems plain common sense to say that there is an invigorating, even mystical, order to the variable shapes of waves as they break, swallows on a summer evening, and weather. Yet for centuries scientists have dismissed such common-sense order. For a long time their attitude made good sense. The traditional task of science has been to simplify nature, expose its underlying logic, and then use that logic as a means of control.

But complex natural phenomena such as weather can't be stripped down, cleaned off, and studied under glass in a laboratory. An individual tree is the result of a vast, shifting set of unique circumstances, a kaleidoscope of influences such as gravity, magnetic fields, soil composition, wind, sun angles, insect hordes, human harvesting, and other trees. An individual wave as it pulsates toward shore is driven and sustained by a beehive of "dynamical" or continuously active forces, far too numerous to determine in detail.

The wave and the tree are dynamical systems, systems whose state changes over time. Systems such as these are multifaceted, complex, and interdependent. They constantly push and pull at themselves to create the sensuous irregularity and unpredictability that is the signature of our physical environment. From the scientific point of view, such irregularity has long been considered a mere messiness obscuring the mechanical, clocklike scientific laws operating beneath. In theory, scientists have believed, the messiness of such systems would be clarified and accurate predictions could be made about their behavior if we could only amass enough information to pinpoint the multitude of their inter-linked causes and effects. Though most people aren't aware of it, many important assumptions we have about nature have been shaped by this scientific idea.

In the twentieth century, we have been overwhelmed by the almost magical ability of science to understand and control our physical environment. This century's dazzling technological progress has led most people to believe that what science doesn't now know about nature it will someday know and that this knowledge will inevitably lead to more and more control. According to this assumption, even the behavior of highly complex dynamical systems will even-tually yield to scientists' formulas and computers. For example, for decades scientists have invested great effort, ingenuity, and technology into studying that vast dynamical system called weather on the assumption—which most of us share—that by improving the quantity and quality of measurements taken on the various factors influencing weather, forecasts would be steadily improved. And it was in weather forecasting that this deep assumption was dramatically overturned.

WIND FROM THE MOSQUITO'S WINGS

In 1961 a Massachusetts Institute of Technology meteorologist, Edward Lorenz, discovered a disturbing fact. He learned that getting more information about such variables as wind speeds, air pressures, humidity, temperature, and sun-spots *won't* help increase the accuracy of a long-range weather forecast. Lorenz ascertained that no matter how much information a meteorologist piled up, his weather prediction would quickly go awry. The reason, he deduced, is that dynamical systems like the weather are composed of so many interacting elements that they are tremendously sensitive to even the tiniest factor. The heat rising from the hood of a car, the wind from the wings of a mosquito in Madagascar, almost anything not included in a meteorologist's measurements can be enough to change the behavior of a weather system. Lorenz's insight meant that in one sense the old assumption was still correct: Complicated dynamical systems are indeed determined by their causes. If we could know all their causes, we could predict what they'd do. But the influences on such a system, Lorenz found, are effectively infinite. As one physicist noted, such systems are so sensitive they can be affected by something as minuscule as the gravitational attraction of an electron on the other end of the universe. So nature is dominated by chaos, but it is not a superficial chaos that theoretically can be reduced to order once we gain enough information. Rather, nature's chaos is profound—because the only way we can ever gain enough information to understand it will be to include the influence of even our attempts to gather the information itself.

With Lorenz's discovery, researchers eagerly plunged into examining all kinds of dynamical systems, from electrical circuits to human brains, and they found new laws. This effort propelled them into an altered view of reality. With impressive speed, scientists moved from their traditional enterprise of studying nature as order to studying nature as chaos, though there was no immediate consensus on how to define what the term chaos might mean.

In mythology and legends, most cultures have wrestled with the idea that order and chaos are a primordial duality. In the Christian tradition, God is described as having moved on the face of the deep (chaos) to bring light (order). The ancient Babylonians told of a mythical hero, Marduk, who slew Tiamat, the cacophonous Mother of All, and transformed her into the order of heaven and earth. In India, Siva, the father of order in heaven, is said to lurk paradoxically in horrible, chaotic places like battlefields and burning grounds of the dead. In the ancient Chinese tradition, daily reality is constantly created and re-created by an oscillation between the light-bringing, ordering principle, yang, and the dark, receptive fullness that contains all matter, yin. The ancient Greeks pitted rational Apollo against libidinous and chaotic Dionysius. The Iroquois peoples of North America cultivated a host of Dionysiuslike spirits, the *gagonsa* or false faces—twisted-looking fright masks that are worn to represent (and purge) psychic and physical disorder. Many tribal peoples around the world include a trickster character among their pantheons, a figure who undercuts order by representing reality's perpetual ironies and deceptions.

Given its quest to simplify nature into a few quantifiable "laws," modern science kept itself largely aloof from such descents into the murky domains of ideas about chaos. However, in the nineteenth century, engineers did discover—to their chagrin—a kind of technological chaos. They realized that energy, or heat, was always lost by their machines, and this led physicists to the notion of *thermodynamic chaos,* a kind of thingless soup that results when hot, organized molecules of directed energy cool off, slow down, and begin to randomly meander into each other. This form of chaos is called "entropy." Nineteenth-century scientific theorists predicted that the universe itself would one day experience a heat-death and end in a whimper of entropy, dissolving galaxies, stars, comets—everything—into a cosmic scale consommé.

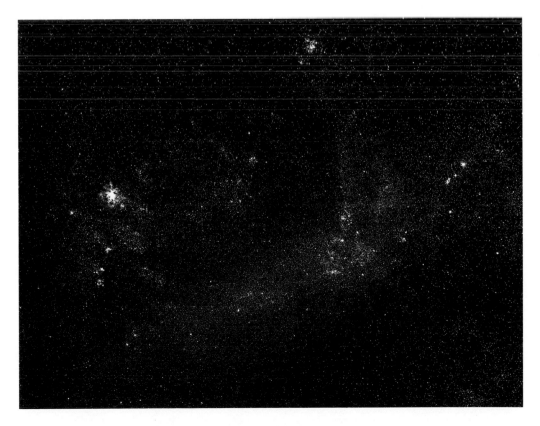

The universe as it explodes and expands leaves behind a fractal imprint of swirling turbulent gases, star fields, and repeating forms. No matter how deeply we peer into space, more detail will always unfold. The photo here is of the Large Magellanic Cloud, a small, irregular companion galaxy to our own found by looking into the region of the Orion Nebula.

SENSITIVE CHAOS

The form of chaos that Lorenz and other scientists discovered in the 1960s and 1970s was perhaps closer to the chaos of the old myths and legends. The chaos, which began to appear like abstract, colorful spirits on computer screens around the world, displayed a wild, haunting order. It was chaos, all right: inherently unpredictable. But as scientists stalked the spirits capering across their screens, they began to uncover a richness in chaos never before imagined.

In order to picture this richness a little, imagine for a moment watching two autumn leaves drop side by side into a stream. The stream, with its leaves, rocks, twigs, bends, and water pressure, is a complex dynamical system that moves in a tortured path through the woods: Straight, smooth stretches are twisted and squeezed into torrents that splash through a labyrinth of boulders, and then slow into quiet pools dammed by downed branches. Moments after the twin leaves fall, they are drawn together by the current into a pool, circling in lazy tandem around a vortex. This doesn't last long, however. Soon the very slight difference in each leaf's position becomes magnified by the water's movement, and the leaves begin to separate. After another swirl, one is swung wide and whisks over the edge of the branch dam, bumping downstream; the other spins slowly into a twig of the dam and is caught there, water backing up against it and rolling in thin, dark ribbons so that the leaf itself now begins to alter, in a small way, the shape and movement of the downstream current.

Chaologists would say that the leaves in this dynamical system exhibited an *extreme sensitivity to their initial conditions.* The very slight difference in their starting points made a very large difference in their fates. Such extreme sensitivity is the hallmark of a chaotic dynamical system. Such systems are highly sensitive because they are always on the move, always changing, never precisely recycling to their initial states. They are like the changing river of time contemplated by the Greek philosopher Heraclitus: You can never step into the river of time twice, Heraclitus said, though it is always the same river. Heraclitus's paradox is also true of a real-life river and is central to chaos. Evidently, even when a complex dynamical system is behaving in a regular and orderly way, at some level the underlying "sensitivity" principle of chaos may be at work subtly separating things, breaking them up. This is not always a bad thing.

Indeed, it is part of the richness of life. For example, in the fetal development of twins with identical DNA, the cells migrating into position to form the twins' brains take different courses and create different patterns of connection. The development of the embryo is a dynamical system, and its extreme sensitivity to initial conditions creates an inherent background chaos which ensures that "identical" twins will never be completely identical.

CHAOS: WINDOW INTO THE WHOLE

One reason that the elements in chaotic dynamical systems are so sensitive to their initial conditions is that these complex systems are subject to *feedback*. For example, through its eddies and turbulence, the water in a stream creates feedback by constantly folding in on itself. Systems fraught with a variety of feedback called "positive feedback" will often undergo revolutionary changes of behavior, such as when a microphone is placed beside a speaker and the microscopic static generated blows up into a deafening screech, or when a tiny grain of ice on a plane wing explodes into a turbulence substantial enough to cause the plane to crash. Systems that change radically through their feedback are said by scientists to be *nonlinear*. As the name implies, they are the opposite of linear systems, which are logical, incremental, and predictable. Linear systems, strictly speaking, are systems that can be described by linear mathematical equations—such things as ballistic missiles and the moon, moving in its orderly orbit around the earth. A spacecraft being nudged by its thruster rockets into a pinpoint touchdown on the lunar surface is a linear system. Like the bursts of the thrusters, small changes in linear systems produce small predictable effects. In nonlinear systems, on the other hand, the folding and refolding of feedback quickly magnifies small changes so that the effect—like the speaker's sudden howl or the small rolling pebble that unleashes an avalanche—seems all out of proportion to the cause. Nonlinear systems behave nonlinearly because they are so webbed with positive feedback that the slightest twitch anywhere may become amplified into an unexpected convulsion or transformation.

The chaologists have learned that in some circumstances nonlinear systems behave in a regular, orderly, cyclical way until something sets them off—a critical point is passed, and suddenly they go chaotic. But then another benchmark may be passed, and they'll return to order again. Imagine, for example, a rock lying on the bed of a stream, a foot or so beneath the surface. When the water is moving normally, the current flows smoothly past the region above the rock, showing no ripple. But after a heavy rain, the speed of the current over the rock suddenly creates an area of turbulence on the water's surface. Then when the current goes back to normal, the surface water courses once again as if the rock wasn't there. Whether chaos rears its head or not depends on the situation. It appears that in dynamical systems chaos and order are different masks the system wears: in some circumstances the system shows one face; in different circumstances it shows another. These systems can appear to be simple or they can appear to be complex; their simplicity and complexity lurk inside each other. Indeed, the chaologists were delighted to find that in modeling dynamical systems, quite simple equations yield results that mimic the unruly dance of chaos. Thus, studying complexity hasn't forced scientists to abandon their faith in nature's simplicity after all—though, to be sure, it has proved a rather strange and uncertain kind of simplicity.

As the chaologists worked, they quickly learned that a dynamical system's

When surfaces crack through the dynamical action of drying, warping, or pressure, they often do so chaotically, creating cascades of self-similar forms at many scales characteristic of fractals. This image, which looks like multicolored dried mud or paint, is actually a layer of polystyrene only one molecule thick compressed and fractured between two sheets of glass.

transition areas—the points at which the system moves from simplicity to complexity, from bright, stable order to the black, impenetrable gyrations of total chaos—were the most interesting places. Inside these transition zones and boundary regions, systems degenerate and emerge in patterns. Though unpredictable in detail, one *can* predict the patterns and ranges of a system's movement. In fact, scientists learned that there are certain repeatable, rough patterns systems seem attracted to as they break down into or emerge from chaos. This discovery delighted scientists because it meant they could still hold on to their scientific reverence for predictability—though now it was a strange and uncertain kind of predictability.

But how do we account for these strange aesthetics of chaos? One rather unexpected answer is "holism." Dynamical systems are sensitive and nonlinear and unpredictable in detail because they are open, either to "outside" influences or to their own subtle internal fluctuations. With the advent of chaos theory, it became impossible to ignore the simple fact that dynamical systems—which, after all, include the most significant processes in our world—don't operate in isolation. The tree that sheds the twin leaves that fall from "outside" into our stream can also be considered an integral part of the dynamical system called the stream. Moreover, within the stream itself, all the elements—from the sharpest bend to the smallest leaf and pebble—constantly interact with each other. In other words, dynamical systems imply a holism in which everything influences, or potentially influences, everything else—because everything is in some sense constantly interacting with everything else. At any moment, the feedback in a dynamical system may amplify some unsuspected "external" or "internal" influence, displaying this holistic interconnection. So paradoxically, the study of chaos is also the study of wholeness.

However, many chaologists would not put it that way. As is undoubtedly fitting to the subject, the debate continues over how to precisely define chaos. Some scientists confine their idea of the chaos phenomenon to the boundary area between stable and purely random behavior. Others prefer to think in terms of degrees of chaos (with randomness at one extreme), arguing that underlying all degrees of chaos is a fundamental holism. But even the holists would agree that the most fertile area of chaos study lies along the ferociously active frontier that has been found to exist between stability and incomprehensible disorder.

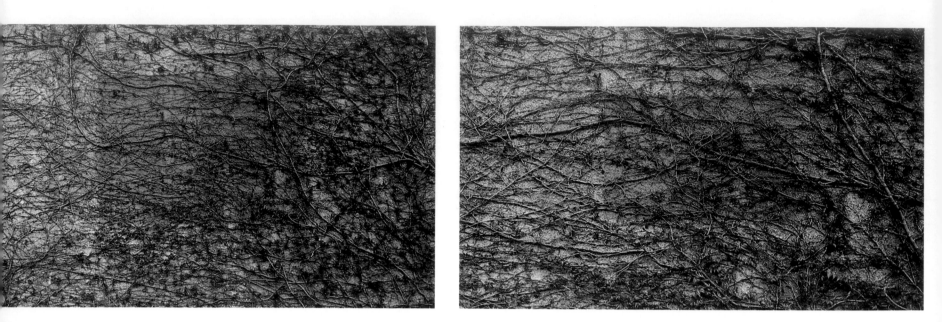

A GEOMETRY OF THE ROUGH

In the 1960s and 1970s an IBM researcher, Benoit Mandelbrot, invented a new geometry, which he called "fractal" geometry, that pushed deeply into this frontier. Mandelbrot coined the term "fractal" to suggest "fractured" and "fractional"—a geometry that focuses on broken, wrinkled, and uneven shapes. Chaos sometimes convulses dynamical systems and sometimes simply resides in the background. Fractal geometry describes the tracks and marks left by the passage of dynamical activity.

We see fractals every day. Trees, mountains, the scattering of autumn leaves in the backyard: all these are fractal patterns, signs of dynamical activity at work. Chaos theory tells the story of the wild things that happen to dynamical systems as they evolve over time; fractal geometry records the images of their movement in space. So a fractal is the fracture left by the jarring of an earthquake or the winding coastline printed with the turbulence of the ocean and erosion; it's the branching structure of a fern which traces the process of its growth; the scrambled edges of ice as it freezes; the spacing of stars in the night sky; the clouds and plumes of pollution spreading out from a power plant. When a chaotic thunderstorm self-organizes into a tornado, it leaves behind it a fractal shape of its destruction. Even the byzantine intricacy of snowflakes is the fractal result of a chaotic process combined with the sixfold symmetry of crystals.

As the camera zooms closer into a vine-covered wall, each magnification reveals new detail which repeats the same patterns discovered at larger scales. According to fractal geometry, this vine is an object that exists between dimensions.

Fractals describe the roughness of the world, its energy, its dynamical changes and transformations. Fractals are images of the way things fold and unfold, feeding back into each other and themselves. The study of fractals has confirmed many of the chaologists' insights into chaos, and has uncovered some unexpected secrets of nature's dynamical movements as well.

One of these secrets is *fractal scaling*. Fractals show similar details on many different scales. Imagine, for example, the rough bark of a tree viewed through successively more powerful magnifications. Each magnification reveals more details of the bark's rugosity. Moreover, in many fractals (such as the tree bark), scaling is accompanied by another corresponding pattern of nature's dynamicism: *self-similarity*. This means that as viewers peer deeper into the fractal image, they notice that the shapes seen at one scale are similar to the shapes seen in the detail at another scale. Perhaps that seems odd. How could systems that are essentially chaotic possess self-similarity on various scales? To understand, consider the weather again.

Seen from space, the earth's weather appears on a vast scale: swirling cloud banks are intermingled with ragged clear regions streaked here and there by more clouds. A snapshot of temperature readings at this planetary scale would show vast hot spots as well as cool areas. Hypothetically, let us suppose that at the global scale we see heavy clouds over North America and that the continent at this moment registers cooler than normal. Suppose we zoom down to the continental scale. It turns out that from this scale the picture looks not unlike what we saw of the planet as a whole. There are quite a few clear areas behind moving fronts of clouds, and here, with more detail in our temperature readings, we learn that some regions of the United States are, in fact, unusually warm. At this scale Colorado seems to be one of those warmer places, so we'll drop down another notch to the state-size scale to have a look. When we tune in to a forecaster on a Denver TV station, the state weather map fills the screen. Once again we see the same variegation we saw at larger scales. There are clouds over Colorado Springs and the temperature is chilly, the forecaster says, but the Aspen area has clear skies and is warm. That forecast might amuse some hikers along the ridges bordering Independence Pass near Aspen. For right now, at their relatively small scale, they are getting rained on. Fortunately, from their vantage point they can see their weather is local. Looking toward a western valley they see the sky above it is clear, and as they move from saddle to saddle along the ridge, they experience a kind of microweather, passing through cool and warm pools of air, intense or slackening bursts of rain, and even a momentary ray or two of sun.

Obviously, the weather at its different scales displays a self-similarity, a fractal structure. One way to explain this is to say that the weather is holistic, which means that between its "parts" (its fronts, patches of rain or snow, high-pressure and low-pressure zones) are other "parts of parts," and "parts of parts of parts" (right down to the shimmers of heat rising from the sweating body of one of the hikers, or the chemical heat generated inside her straining muscle tissue). The result is that when all these "parts" and "parts of parts" start feeding into each other, they can generate images (such as weather maps) whose patterns have scaling detail. These patterns illustrate the fact that the system's whole movement takes place continuously at every scale.

In the abstract, Euclidian world, scale isn't important, and magnifying spheres, triangles, squares, or lines won't yield much new information about the object at hand. In the fractal world, there are wrinkles and crinkles, sometimes infinite detail, and more and more information the deeper we go. In the Euclidian world the observer moves in discontinuous jumps from the one-dimensional line to the two-dimensional square to the three-dimensional cube. In the fractal world, dimensions are tangled up like a ball of twine, and objects are neither two dimensions nor three but somewhere in between. In fact, fractal geometry has come to be known as a *geometry between dimensions.* Depending on its wrinkling or fragmentation, a fractal object may be any one of an infinite number of possible fractional dimensions.

Fractal images have led to a growing contemplation of our reality as a place made up of folded worlds within self-similar worlds—that is, of worlds folded in between dimensions. Bend down to look at a moss-covered rock and you see a miniature mountain range covered with trees, a microcosm of our larger landscape. But if it's true that everything on the planet has evolved through intense interaction with everything else, then these self-similar images of holism we see around us should perhaps not be surprising. The fingers on our hands are self-similar to the wings of a hummingbird and the fins of a whale. After all, we all evolved inside the same holistic dynamical system called life.

COMPUTER AS MICROSCOPE

When scientists and mathematicians began to work with fractal geometry they learned to their amazement that they could generate intricate fractal forms on their computer screens with fairly simple nonlinear formulas. These formulas have feedback terms: the result of a calculation is input back into the equation, and the equation is run again. Continually rerunning an equation inputted back into itself is a process scientists call *iterating the equation.* This leads to fantastically complex, sometimes eerily beautiful structures that display fractal self-similarity. One of the most well known of these structures can be created by using the computer to iterate an equation involving a particular set of numbers named after Benoit Mandelbrot, who was the first to uncover their beauty.

Elegant and fiendishly clever self-similarity appears along the boundaries of the Mandelbrot set, making this purely mathematical construct an emblem of real-world processes of chaos where fractal self-similarity exists at the edges of waves, in fracture zones, and along weather fronts. Scientists now regularly

An explosion of fractal self-similarity at different scales occurs in the boundary area of the Mandelbrot set. The set is actually an infinite cluster of numbers on the complex number plane constituting what has been described as "the world's most complex mathematical object." To generate this particular spiderlike image from the boundary area of the set involved millions of mathematical calculations by the National Aeronautics and Space Administration's massive parallel processor.

use iterated fractal formulas to model the unfolding and gyrations of real dynamical systems such as turbulent flows of water or gas.

It would be hard to overestimate the role the computer has played in the revolutions of fractals and chaos. Without the calculating power needed to iterate equations millions of times, the revolution simply would not have been possible. The high-speed, number-crunching computer became to the study of complex dynamical systems what the microscope was to the study of microbes, the particle accelerator to the study of subatomic structure, and the telescope to the study of deep space. The computer brought phenomena into focus that scientists had never seen before. The computer's power to make vivid images out of mathematical models led to a growing appreciation of the complex beauty of chaos. One surprising result has been to draw two cultures together that have been separated by hundreds, if not thousands, of years.

DISCOVERING A NEW (AND OLD) AESTHETIC

Chaos theory and fractal geometry extend science's ability to do what it has always done: find order beneath confusion. However, the order of chaos imposes a definite limit on our ability. With the use of computers, scientists can see chaos, can understand its laws, but ultimately can't predict or exert control over it. The uncertainty built into chaos theory and fractal geometry echoes two earlier scientific discoveries of this century: the fundamental uncertainty that Gödel's theorem found skulking inside mathematics and the array of essential atomic uncertainties and paradoxes unearthed by quantum mechanics. Science, in this century, seems destined to learn about nature's intention to remain behind a veil, always slipping just beyond our understanding, imposing a subtle order.

Artists have always exploited and valued what might be called "the order that lies in uncertainty." The British Romantic poet John Keats admired what he called "Negative Capability," the ability to be "in uncertainties, mysteries, doubts." He claimed that this capacity was key to the artist's creative power. Leonardo da Vinci insisted that "that painter who has no doubts will achieve little," and he advised fellow artists to seek out inspirations for their paintings in the stains on walls. Artists have perennially discovered in the doubt, uncertainty, and haphazard of life a harmony that goes straight to the essence of being. Whatever it is that the painter, poet, or musician depicts—whether abstract or realistic—the artist's final product implies worlds within worlds. Within art there is always something more there than meets the eye, the mind, or the ear. Because of this ability to intimate worlds within worlds, art has always been fractal. The science of chaos is helping to newly define an aesthetic that has always lain beneath the changing artistic ideas of different periods, cultures, and schools.

Many contemporary artists, like Connecticut-based landscape painter Margaret Grimes, immediately recognized in chaos theory a deep connection to their personal artistic orientation to the world. Says Grimes: "These ideas confirmed mathematically something that I had already perceived experientially through observations of nature. The theories thus had great resonance, as of a truth one has always known but has not known how to express."

New York painter Nachume Miller participated in a 1989 art exhibit on chaos after he realized that chaos theory applied not only to the subjects he painted but also to the artistic process by which a painting comes into being: "The way I go into certain processes is fairly chaotic, not very clear to me. You respond to a chain of events that happen when you work. You first have to create a drama on the canvas that is very disturbing. You don't actually know what it is. You don't even like it, and then through looking at it more, getting more familiar with what's going on, you get some more clarity."

Oregon photographer Joseph Cantrell describes a similar process in his own work: "The order is out there in so many planes for which we either have no perception or have been trained not to see it. I shoot for the surprise. Very often I get it in some of the most prosaic subjects. There's a state you can get into when photography is going well where you lose yourself. At the end of it you've been somewhere that's pretty wonderful but you can't remember the details until you see the final result." The results are a fractal record of his interaction with his subjects, which are usually fractal objects themselves such as ferns, volcanos, and turbulent water.

"A collision of forces that occur when boundaries are eliminated . . . these dark and turbulent paintings [have the] power to evoke many different realms," reads the catalog to Nachume Miller's 1988 show at the Museum of Modern Art. Miller himself says, "Looking at my work, you could see a seascape, a microscopic cosmos . . . it could be the Milky Way. It's finding out things about your mind." After becoming acquainted with the scientific ideas of chaos theory, Miller now also sees his works as repositories of self-similar forms. He calls this particular painting of oil and wax a "landscape" because of the vague sense of earth and sky and the way the light breaks through the turbulence.

This new (and very old) aesthetic brought out by chaos might be described as follows:

It is holistic: a harmony in which everything is understood to affect everything else. In mathematical fractals and also natural fractals, the holism appears as self-similarity, evidence of a holistic feedback process. In art, self-similarity—which can come in infinite variety—is not created by a slavish permutation of some form at different scales. Rather, it is closer to the self-similarity seen when we compare a human hand to a hummingbird's wing to a shark's fin, and to a branch of a tree. It is the artist's task to find and express this significant relation between forms and qualities that are simultaneously self-similar and self-different so as to create an artwork that allows us to glimpse the holistic nature of our universe and our being in it. Miller says of his own work, "If you take a fracture of my stuff, typical of the patterns I use, in principle it is very much like the totality of the picture. It will have the same kind of logic as the whole."

Miller insists that the artist isn't trying to "represent" nature. "Instead of illustrating nature the pictures want to work like nature." They should be "like" life-forms, in other words—and an essential feature of life-forms is that each in its own fractal way reflects the dynamical system of nature as a whole. The holistic element is an essential feature of this new (old) aesthetic appreciation. It's why when you see a colored pebble gleaming on a beach among a jumble of others and take it home, it may not look as lovely on the shelf as it did in the natural chaos where you found it. To cut a path through the woods or a highway through the jungle is to recognize that through your gesture the entire landscape has become altered. Chaos affirms that individual details matter. Artists know that like the sensitivity of a chaotic dynamical system, a change in one small part of a painting or a poem may destroy or transform the work.

The holism of the new aesthetic also brings out a new (and very old) relationship between the observer and the object observed. The Greek roots of the word aesthetic suggest that an aesthetic experience involves a transformation which takes place in both the object and its observer. Science has traditionally assumed that the observer could stand off and be "objective" about what he observed. Chaos theory has revealed, however, that observers are inextricably

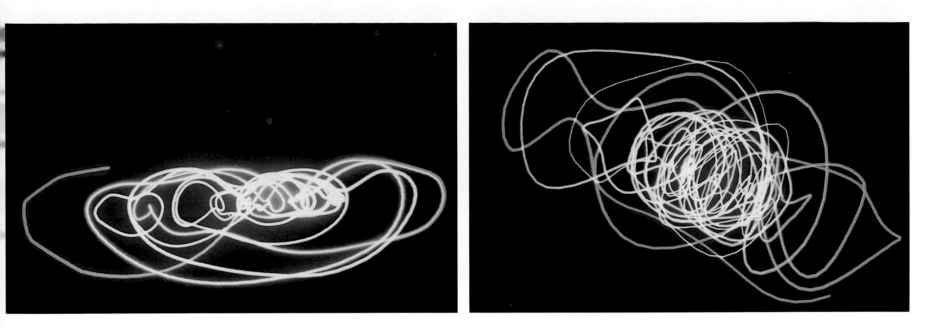

The brain's "strange attractors" show up in these two plots of the electroencephalographic data collected from a woman with her eyes closed resting, left, and performing a seven-step arithmetical problem, right. One of the discoveries of chaos theory is that the brain is in fact organized by chaos. In this experiment, neuroscientist Paul Rapp found that the chaotic activity of the brain was attracted to specific regions of the plotting space in each of the two states. Thus there is a strange attractor for the mind when it is resting, and another for a mind doing a math problem.

part of dynamical systems, something artists have always understood. With chaos theory, it is no longer possible to assume that an observer can blithely analyze an object or process into its constituent parts (an assumption that in science has been called "reductionism"), because "parts" are dynamic and therefore have unpredictable effects. Grimes sums up the new aesthetic when she says that its vision is of "structure/evolution, stability/spontaneity—infinite generation: The pattern we perceive is composed of an endless variety of interwoven patterns. The concepts of order and chaos reflect the absolute relationship of the whole and yet the infinite significance of each part whether the part is an element or an action or a process."

The study of chaos has clearly sensitized scientists to the aesthetic experience of art. Paul Rapp, neuroscientist at the Medical College of Pennsylvania, admits that the forms he's plotted on his computer don't compare with the "worst Monet." But he describes with the enthusiasm of an artist his reaction to these fractal plots—mathematical representations of a human brain thinking. His video pictures of electroencephalographic (EEG) data show that while our brain's electrical activity is chaotic and unpredictable, it has a hidden order in that it is attracted to a certain region of the plot space. He says of the discovery of these fractal *strange attractors* in the brain: "The emotional impact of electroencephalographic images is, for me, rather considerable. For the first time we are able to see the changes in the geometry of EEG activity that occur as the result of human cognitive activity. Before these attractors had been constructed, I didn't know what to expect. I expected to see something very boring that did not significantly change as the subject began to think. The moment these structures flooded onto the screen and began to rotate, I knew that I was seeing something very extraordinary."

Scott Burns, an associate professor of general engineering at the University of Illinois at Urbana-Champaign, says he's seen the images he generates of mathematical chaos excite awe in viewers. "I have a colleague who said, 'Boy, this is sure a basis for a belief in God.' I wouldn't go that far, but I would say it's certainly a basis for reverence of nature."

To get a feel for how the aesthetics of chaos is bringing the two cultures of science and art together, compare statements by Mario Markus, a physicist at the prestigious Max Planck Institute in Dortmund, Germany, and Eve Laramée, a New York-based sculptor.

In his lab Markus generates brooding, monstrously graceful fractal images of an important set of equations used to model turbulence. The control Markus has over variables such as which equations to portray, which mathematical values to start with, what colors to assign to the values, what scaling and intensity level to use is, he says, like the control that the photographer exercises over his subject matter and is not just a mechanical pressing of the button on the computer. He argues, "The particular choices made by one person, as compared to those made by others, allow us to speak of a personal, recognizable 'style.' Truly one can say that equations can be considered here as new types of painting brushes."

Laramée creates ancient-looking constructions out of copper, salt, and water. Once she has installed one of her artworks in the gallery, the salt dissolves and begins to eat intricate, ageless fractal shapes into the copper so that the piece evolves over time. While Markus strives to insert himself into his equations and exert some control over the chaos that automatically unfolds, Laramée strives to take herself out of the process and let the inherent chaos roll in. She says, "There is a point where I 'remove' the hand of the artist, and allow nature to take over and finish the work." Thus the new aesthetic created by chaos ensnares both artists and scientists, both observer and observed. The so-called objective/subjective wall that for centuries has divided scientists and artists in their approach to nature is now being shattered from both sides.

From space we immediately see that our planet is fractal. The red areas in this satellite shot of the prominently fractured region around the Ala River on the Nigerian-Cameroon border in Africa indicate vegetation on the mountainsides. The valleys and plains were colored blue-green by the computer, indicating heavy cropping and land use. The highly fractured geology of the region displays dendritic scaling (branching forms of many different sizes that make it look like a network of blood vessels). This is a signature of fractals. The intricate fractal design of this region records the dynamic action of geological forces. The sinuous black line of the Ala River follows the fault lines, as if to underline the fractal pattern.

A UNIVERSE FULL OF CHAOS AND FRACTALS

Perceptions change almost hourly as artistic and scientific investigators peer through the windows of fractals and chaos to discover meaningful patterns of uncertainty everywhere: The surfaces of some viruses are now known to be fractal. Fractal rhythms and distinct fractal signatures have been found in dopamine and serotonin receptors in the brain, and in enzymes. Fractal geometry is being used to describe the percolation of oil through rock formations. Composers are creating fractal music; programmers are studying the effect of chaos on computer networks; chemists are applying fractals to the creation of polymers and ceramic materials; economists are locating a strange attractor underneath the fluctuations of the Standard and Poors Index; ecologists are using the principles of self-organizing chaos to reconstruct lost habitats; nonlinear models have been made of the international arms race. One enterprising novelist has turned the idea of strange attractors into a science fiction story equating chaos with immortality.

Many other views of fractals and chaos can be seen in the chapters that follow.

A PLANET OF LIVING FRACTALS

•

If the eye attempts to follow

the flight of a gaudy butterfly,

it is arrested by some strange

tree or fruit; if watching an

insect, one forgets it in the

strange flower it is crawling

over; if turning to admire the

splendour of the scenery, the

individual character of the

foreground fixes the attention.

The mind is a chaos of

delight . . .

—CHARLES DARWIN,

writing home from his

Beagle voyage on his

impressions of the

Brazilian tropical rain

forest.

•

Hike into a forest and you are surrounded by fractals. The inexhaustible detail of the living world (with its worlds within worlds) provides inspiration for photographers, painters, and seekers of spiritual solace: the rugged whorls of bark, the recurring branching of trees, the erratic path of a rabbit bursting from underfoot into the brush, and the fractal pattern in the cacophonous call of peepers on a spring night.

The landscape is the crucible in which living forms have evolved, and since the landscape crackles with fractals, the forms bred there are fractal as well. Living creatures, from trees to beetles to whales, have shapes and behaviors that provide a fractal record of the dynamical forces (the endless feedback) that act upon them and within them, forces that have continually caused them to evolve new niches in which to live. In his *Boston Globe* newspaper column, physicist and science writer Chet Raymo declared after seeing a museum ex-

This photograph by Lawrence Hudetz of Oregon's Columbia Gorge is alive with fractal shapes that result from forms continuously evolving together.

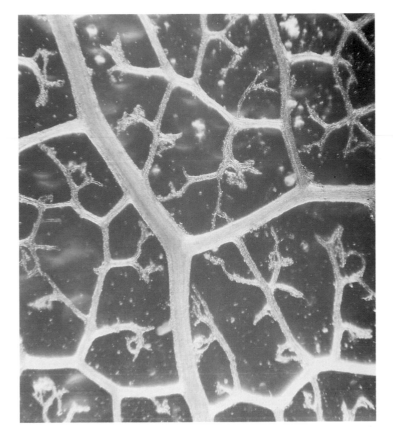

If we were to zoom in, we would see that the fractal shapes in Columbia Gorge carry down to the microscopic scale shown by this photograph of leaf veins. This image was made by the late Lewis Wolberg, a psychiatrist fascinated by the aesthetics of science, nature, and art. Wolberg said, in his book *Mico-Art, Art Images in a Hidden World:* "Why do the representations of some artists so often resemble microscopic structures?" He answered, in part, that artists "may be responding to the same interacting processes that operate in all of creation. As Emerson expressed it in his essay *Nature:* 'Compound it how she will, star, sand, fire, water, tree, man, it is still one stuff, and betrays the same properties.' "

hibition of beetles, "Darwinian explanations are reasonable enough, but . . . the spectacular variability of beetles suggests that nature is infected by . . . a sheer lunatic exuberance for diversity, a manic propensity to try any damn thing that looks good or works."

The riotous beauty and dreamlike strangeness of nature provided a chief inspiration for Charles Darwin as he struggled to develop a coherent theory of evolution. Psychologist Howard Gruber, who has done a lengthy study of how Darwin arrived at his theory, says, "The meaning of his whole creative life work is saturated with . . . duality. . . .

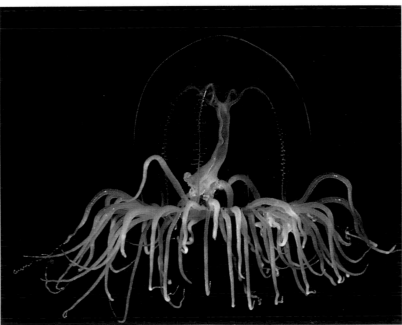

The primordial shape of the jellyfish is a sinuous record of the dynamical forces of the flowing water in which the creature evolved.

It would take some doing to calculate the fractal dimension of this mottled, warty-shaped inhabitant of tropical waters. The frogfish can change colors to match its surroundings and become almost indistinguishable from an algae-encrusted rock. Nature makes cunning use of symmetry and fractal irregularity to create its organic forms.

On the one hand, he wanted to face squarely the entire panorama of changeful organic nature in its amazing variety, its numberless and beautiful contrivances, and its disturbing irregularity and imperfections. On the other hand, he was imbued with the spirit of Newtonian science and hoped to find in this shimmering network a few simple laws that might explain the whole movement of nature." Darwin concludes his landmark *Origin of Species* with a striking metaphor of nature as "the tangled bank," reveling in what Gruber calls "the spectacle of complexity itself." Indeed, the pattern—the image—that gave Darwin his essential insight into how evolution works was a classic fractal: He conceived of the evolving forms of nature as an irregularly branching tree.

Examining Darwin's notebooks, Gruber carefully tracked Darwin's creative process to the moment when this image emerged in his thought. Gruber initially expected Darwin's mental processes on evolution would be "fine, clean, direct," but soon found that they were "tortuous, tentative, enormously complex." Gruber realized that "Darwin's picture of nature as an irregularly branching tree attributed to nature some of the characteristics I saw in his thinking."

According to Gruber, after considerable mental bifurcation Darwin reached a point where he drew in his notebook three tree diagrams which captured his insight that all creatures are related to one another through a process of branching pushed forward by natural selection. Darwin had found a simple law that could explain life's breathtaking complexity.

Through the ages artists have been driven by a desire to capture life's simultaneous complexity and simplicity in a single image or work. Some artists have created simple images with hugely complex overtones; others have spun out complex images that imply a simple order beneath. Artistic "truth" seems to involve presentation of a *dynamic balance* between these two opposites. Darwin's admiration for complexity and his belief in the Newtonian model of simple natural laws brought him an important step toward the artist's aesthetic (sense of harmony and dissonance), but in the end the emphasis of evolutionary theory fell on the simplicity side of the equation—on scientific law. Many of the scientists of chaos (though certainly not all) now seem bent on readjusting the balance. Accordingly, they are proclaiming a new dynamic that emphasizes how complexity can be wrought from simple rules while at the same time revealing a challenging new perception that the laws of complexity will forever prevent the kind of simple predictability and control over nature implied by the clockwork Newtonian model of the world that Darwin had admired.

This apparent piece of modern sculpture is in fact the fractal shape of a ginger root, one of nature's many living irregularities.

Using simple mathematical rules, chaologists can now model complex dynamical systems, formulating rules to mimic on a computer such natural phenomena as the flocking pattern of birds flying to a roosting spot and the growing branch and leaf forms of specific flowers and trees. Chaos theory and fractal

This stained cross section of cells in a cucumber bears a curious resemblance to the purely mathematical fractal pattern generated on a computer. Michael Barnsley has calculated and graphically represented here the values in a portion of the boundary of the Mandelbrot set, an infinite collection of numbers found on the complex number plane.

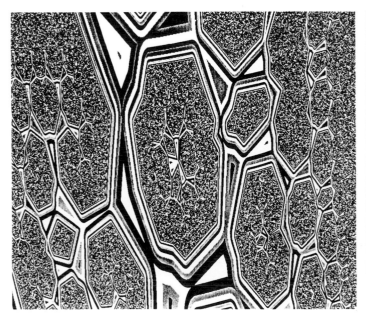

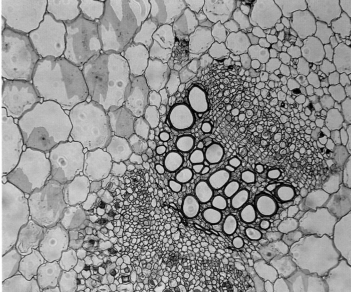

Notice the fractal patterns on the symmetrical starfish and the fractal composition created by their bodies washed up on the rocks. They are beneficiaries and victims of nature's dynamical action.

geometry have opened up undreamed of correspondences between the abstract mental realm of mathematics and the movements and shapes of our planet's myriad organisms. The seemingly endless niches in nature, for example, can now be perceived as an analogue for the intricate complexity which fractal geometers have found in the nooks and crannies of the Mandelbrot set. Indeed, the idea of niche itself can now be understood as a fractal idea.

Niche means a corner or space. Biologists have traditionally used the word to signify the little empty corner of an ecosystem that an organism evolves to fill; a niche presents an opportunity for evolution. If one species of cormorant nests on high cliffs with broad ledges and eats a certain kind of diet, another species will evolve with special characteristics that allow it to nest lower down on narrow ledges and eat a slightly different diet—so the two species occupy different niches. In this traditional view, nature abhors a vacuum and will evolve new forms to fill it. But, in fact, the situation is considerably more subtle. An organism creates the niche it occupies as much as it is *created by* the existence of an unexploited region of the ecosystem. New spaces or niches constantly come into being, unfolded by the total activity of organisms. When a species dies out, the fold (or niche) smooths over or is further crumpled into other folds. The great biological diversity on the planet is a sign that nature is continually rippling with new and related niches. It is like the surface of the sea wrinkling in the wind.

The constant crumpling of reality that we see in evolution takes place over millennia as species emerge and pass away, creating new landscapes, new environments, and new opportunities for new species. The old scientific concept of the "balance of nature" is quietly being replaced by a new concept of the dynamic, creative, and marvelously diversified "chaos of nature."

The eighteenth-century British satirist Jonathan Swift took a humorous view of nature's scaling:

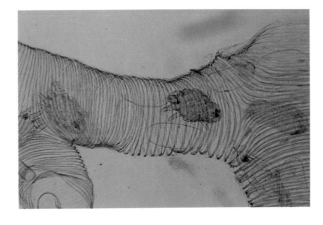

So, Nat'ralists observe, a Flea
Hath smaller Fleas that on him pray,
And these have smaller yet to bite 'm,
And so proceed, *ad infinitum*.

   Probably Swift would have been suitably amused at this photograph which shows mites inside the trachea of a bee. Swift was correct that life is built on the principle that evolutionary activity creates worlds within worlds, all moving, changing, feeding back into each other from small scale to larger scale, back to small scale.

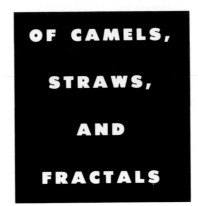

OF CAMELS, STRAWS, AND FRACTALS

•

To call a general differential equation "nonlinear" is rather like calling zoology "nonpachydermology." But you see, we live in a world which for centuries acted as if the only animal in existence was the elephant.

      —Ian Stewart, mathematician, in *Does God Play Dice: The Mathematics of Chaos.*

•

Most things in nature are like the camel that had one too many straws on its back. Continental plates push against each other for centuries and nothing happens—then suddenly, an earthquake. The boss, who relentlessly drives himself and everybody else and seems to have boundless energy, drops dead of a heart attack. A new insect is introduced accidentally into the environment, and at first its population explodes, a few years later collapses, then is stable awhile before exploding again.

Regularity, abrupt changes, and discontinuities are primary features of life. Scientists call such jagged behavior "nonlinear," and the name is a clue as to how they feel about it—or felt about it until recently. Nonlinear means not

This portrait of a nonlinear equation was created by an unusual collaboration between an artist and a scientist. Gottfried Mayer-Kress of the Sante Fe Institute is one of the world's experts on nonlinear systems. In the early 1980s he saw a potential for artists in the dynamics he was plotting, "but I also knew that I—as a scientist without artistic talent or training—was not the person who could uncover these other, nonscience layers of chaotic structures." He eventually teamed up with graphic designer Jenifer Bacon who was entranced by the images Mayer-Kress showed her: "The chaos images on the computer were like the land and sky in which I could paint and interpret what I wanted,"

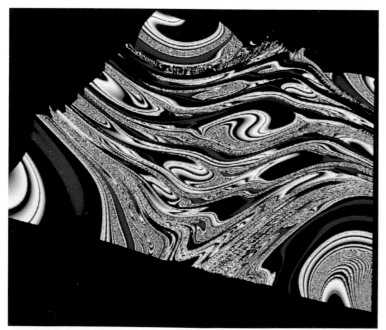

she says. "There was something intriguing about the structure of the images. They seemed to flow and move like liquid or atmosphere."

Sometimes Bacon takes off from the science to transform the images into her own landscapes. Other times, she uses her aesthetic sensibilities to heighten the visual impact of the scientific information in the plot. The image here is a faithful, though artistic, portrait of a nonlinear equation showing in graphic form the huge range of behavior possible depending on the starting values applied to the equation. The "island" in the lower left, for example, indicates values where the equation shows periodic behavior. This equation has been used to model the behavior of subatomic particles.

linear, and the implication is that linearity is the preferred state. It all has to do with equations.

Equations can be thought of as science's similes and metaphors. When physicists, chemists, or biologists use equations to model real processes in nature, they assume that the way an equation unfolds is like the unfolding of the real process that the equation models. The values in linear equations change in an orderly way, by steps and proportions, so the cause and effect processes of nature in the linear world are lawful and orderly, just as Isaac Newton described them in his work on celestial mechanics. A nonlinear equation is strikingly different and provides a strikingly contrary picture of nature.

While solving linear equations is a simple matter of slotting in quantities and calculating the equation's terms to a result, nonlinear equations must be solved by *iterating,* or recycling, the end result of the equation to see whether processing the equation pushes that end value toward a stable number, a periodically returning number, or a number that fluctuates randomly. This suggests that the cause and effect processes of nature described by nonlinear equations themselves involve some kind of dynamic recycling that leads to stability, periodicity, or chaos. If mathematicians solve a linear equation with one starting value and then solve it again with a closely related starting value, the end results of the two calculations will remain close to each other. If these same mathematicians try plugging two similar starting values into a nonlinear equation, the results of the two calculations might be close or they might be shockingly far apart. While a linear equation will behave the same way almost no matter what values are slotted into it, a nonlinear equation is exceedingly sensitive to its starting conditions. With a linear equation, when you've solved for one value, you have a good idea of how the equation will behave when you solve for any value. With a nonlinear equation, you have no such assurance. So, while linear and nonlinear equations both describe the relationship of causes to effects, metaphorically speaking, they seem to describe the causal behaviors of nature on entirely different planets.

For a long time scientists could formulate nonlinear equations that modeled some of nature's complex processes, but couldn't solve them. Unable to solve them, scientists did the natural human thing and *linearized* all the nonlinear,

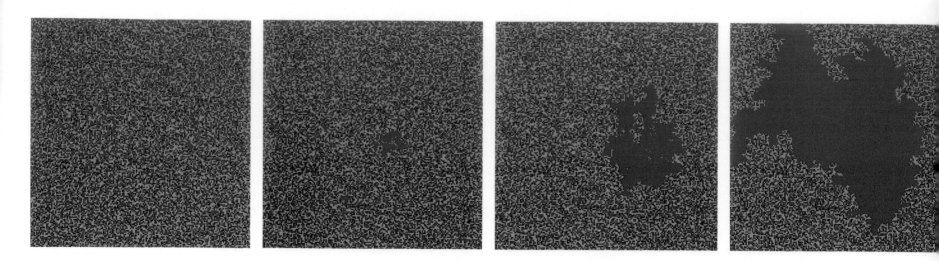

Scientists at the Brookhaven National Laboratory in New York have been studying the nonlinear dynamics of sandpiles. Like the camel that collapses under one straw too many, a sandpile added to one grain at a time will reach a critical mass and then suddenly collapse. Per Bak, Kan Chen, Michael Creutz, and a team at Brookhaven have constructed a computer model that plots a sandpile avalanche. The first frame plots a sandpile that appears stable in the sense that nothing is sliding. The colors red, blue, and green indicate different levels of stability for individual grains. Black indicates open spaces. The muddy red area in the second, third, and fourth frames show a spreading avalanche that occurs when one grain more is added to the pile, setting off a sudden change in its behavior typical of nonlinear systems. Notice the outline of the avalanche wave is fractal. The yellow spots indicate grains that are still rolling. In the real sandpiles studied by scientists at IBM in Yorktown Heights, New York, avalanches of sand drizzle down the side of the pile in fractal patterns.

Both the real and the computer experiments have shown that sandpiles "perpetually organize themselves to a critical state in which a minor event starts a chain reaction that can lead to a catastrophe," report Bak and Chen. Avalanches tend to maintain the pile at the critical state, so even though the pile rises and falls as grains are added and avalanches occur, as a whole the pile is always evolving toward its most unstable state, a process Bak and his colleagues have dubbed "self-organized criticality." Says Bak, "The geometric description [of fractals] does not explain anything in itself. One has to understand the dynamical origin of fractal structures. I see our idea of self-organized criticality as a contribution in that direction. . . . The dynamics of self-organized critical systems [are] 'at the edge of chaos.' . . . We believe that the fractal structure of nature indicates that nature is turned to the edge of chaos."

tumultuous phenomena they could, such as heat flow—and then neatly dismissed the behavior of any "messy" natural phenomena they couldn't linearize. Linearizing involves throwing away the awkward terms in the nonlinear equation (the terms that involve feedback), using instead a series of approximations to model the process at hand.

At the turn of the twentieth century, physicists used linear approximations to calculate and predict the movement of planets and satellites in their orbits. It was a highly successful procedure. Then the great French scientist Henri Poincaré took up the challenge of solving a nonlinear equation that involved the feedback of gravitational effects that are produced when more than two celestial bodies interact with one another as they move. The calculations were immensely complex, but Poincaré soon discovered that chaos is present in the very celestial mechanics that linear science had long trumpeted as the model of nature's simple laws. Poincaré was stymied, however, by the strangeness of the results and the immensity of the calculations; so he abandoned the nonlinear approach. Then computers came along, which could crank out the millions of iterations necessary to solve a nonlinear equation. Almost overnight scientists began to explore nonlinear equations as a potent and revolutionary mathematical metaphor for nature.

The shapes and figures that appear on computer screens when scientists iterate nonlinear formulas are fractal shapes—mirrors of underlying tempestuous dynamics.

In a nonlinear world, small effects can have large and unexpected consequences. A dynamical system may seem stable until it reaches a critical juncture, and then some seemingly minor occurrence pushes it over the edge to a new state. In the frame at the right an iceberg suddenly breaks off, and "calves" from a glacier. Polar ice shows fractal scaling: Ice fragments are mirrors, at different scales, of each other. In the frame at the left, an avalanche, showing its fractal edge, rolls down a mountainside in British Columbia. The avalanche could have been set off by something as small as an echo or a change in temperature. Nonlinear formulas would be needed to describe the sudden changes shown here.

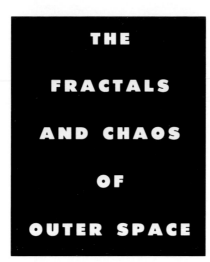

THE FRACTALS AND CHAOS OF OUTER SPACE

•

It has lasted for nearly a

billion years without flying

off. If I had to bet, I'd say it

wasn't going to—but we can't

rule it out.

> —JACK WISDOM, MIT
>
> physicist, speaking
>
> about the planet Pluto.

•

pace probes and flybys, cameras and telescopes equipped with sophisticated X-ray and ultraviolet sensors, unmanned landings and manned lunar expeditions—all have combined to bring us spectacular views of our solar system as a place full of bubbling, freezing, oozing, shattering activity. The swirling cyclone of gases that makes up Jupiter's giant eye is only one instance of the dynamical chaotic forces that operate among the planets and in the deep space beyond. Here collapsing neutron stars spin at frantic rates, supernovas slowly explode in shock waves that trigger the birth of new stars, suns—spinning balls of turbulence—spew out magnetic storms across millions

A cauldron of chaos, the Orion Nebula is a crucible or womb of dust and gas in which stars are born.

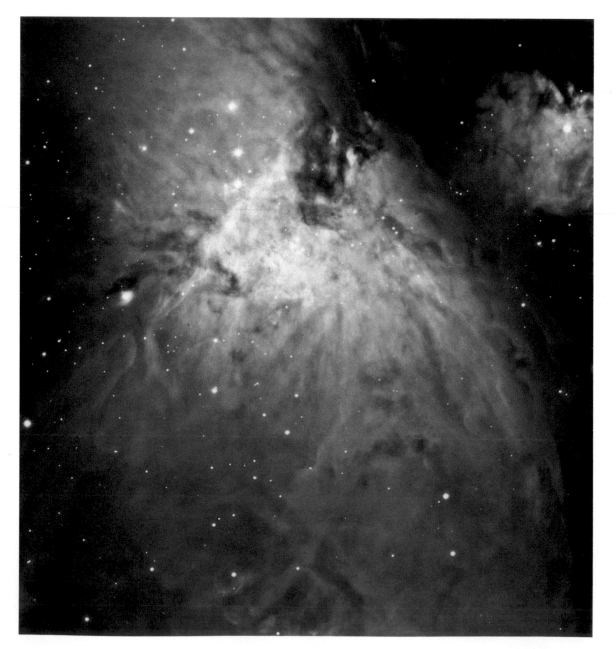

of miles, and black holes chew up passing energy.

It's hard to imagine that not very long ago our solar system was considered the very paragon of nature's order. Ever since the late seventeenth century, Isaac Newton's great theory of "celestial mechanics" has enabled scientists to predict with amazing precision such astronomical events as planetary cycles, solar eclipses, and comet returns. Newton's laws of gravitational attraction increased scientists' understanding immeasurably, even leading to the discovery of new planets. By the eighteenth century instrument makers were using gears and pendulums to fashion "orreries," sophisticated clocklike instruments that kept the repetitive orbital time of the planets as they circled in the solar system.

Then around the turn of the century the great French physicist-mathematician Henri Poincaré encountered a disturbing glitch in Newton's celestial mechanics. The equations traditionally used for calculating the gravitational attraction of celestial bodies work wonderfully when the planets are taken two at a time. But when the effect of a third object is added, the equations become unsolvable. Physicists had traditionally gotten around this so-called three-body problem by using what are called "linear approximations"—for most practical purposes a fine solution. Poincaré, however, decided to work theoretically on the problem by adding a term to the equations that would represent the feedback caused by the presence of the third body. This term made the equations "*non*linear" and gave Poincaré considerable distress. Nonlinear equations be-

Orreries like this one made for Harvard College in 1767, by Benjamin Martin of London, represent science's view of a universe under the strict guidance of "celestial mechanics." That view has been challenged by the recent scientific realization that this clockwork system contains traces of chaos.

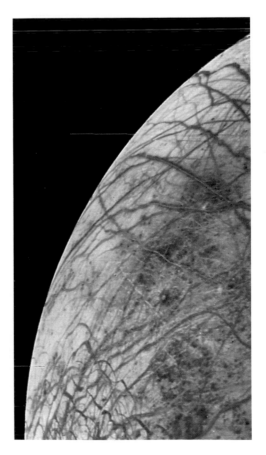

Like a bloodshot eye, Jupiter's moon Europa displays a fractal pattern evidencing the complex dynamics going on at its surface. The chaotic web of red lines—and lines within lines—are fractures in the 100-kilometer-thick crust of ice covering the planet. The fractures are being filled with materials from the planet's interior.

have erratically as terms are rapidly multiplied by the equation's dynamics. Poincaré's solutions suggested that in some orbits, the presence of a third body might cause a planet to gyrate, weave, even fly off.

The strange behavior of the equations meant that the eternal clockwork regulating the planets' orbits might come unexpectedly unsprung. Poincaré soon abandoned the calculations, lamenting that his results were "so bizarre that I cannot bear to contemplate them." In his wake, scientists carried on with their linear approximations, ignoring Poincaré's nonlinear feedback as so much experimental "noise."

But what is noise for one era of science may, in a later era, become the drumroll to a new reality. In the past twenty years, "chaologists" have returned to contemplate the discovery Poincaré abandoned—and increasingly found evidence of chaos in the celestial machine. Chaologists have, for example, ascertained that gaps in the asteroid belt between Mars and Jupiter are caused by Mars' gravitational attraction, which though small when compared with the attraction of Jupiter, is large enough to create regions so chaotic that no asteroids can reside in them. Jack Wisdom, an MIT physicist specializing in celestial mechanics, speculates that some asteroids which wandered into the gaps may have been hurled toward Earth, where they crashed as meteorites.

Scientists have also found chaos in the tumbling of Saturn's deflated rugby ball–shaped moon, Hyperion.

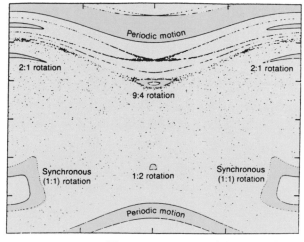

A plot of Hyperion's orbit captures the strange, wobbling behavior of this 120-mile-long satellite as it whirls around its mother planet. Hyperion's rotation behaves regularly and predictably only in spurts, which show up in the plot as yellow islands of order amidst a blue sea of chaos. In some of these islands of order Hyperion rotates twice every time it orbits Saturn; in others it spins nine times for every four orbits; there are also bands of regular or "periodic motion."

Hyperion's tumbling behavior has been confirmed by observation and is so unpredictable that chaologist Wisdom says, "Even if it had been possible to

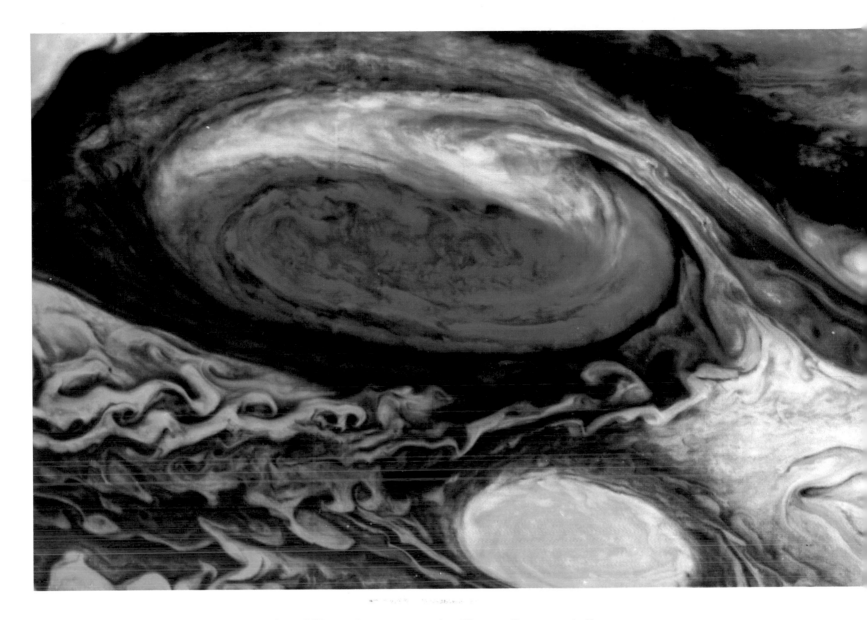

determine the orientation and spin of Hyperion to ten significant figures at the time of the *Voyager 1* encounter, it would not have been possible to predict its orientation less than two years later when *Voyager 2* arrived." The self-similarity represented by the islands of order in the plot are fractal: fingerprints showing that a chaotic dynamical system is at work.

Chaologists now think that the orbit (not spin) of the planet Pluto may also partially occupy a chaotic region. They think it is possible Pluto could one day suddenly lurch off into a new orbit.

From plots of planetary uncertainties to the shapes and features of celestial objects, fractal patterns exist across the cosmos. The pattern of distribution of craters on the lunar surface is fractal as is the scattered pattern of galaxies in the universe. In the latter case, star clusters contain gaps, and in the gaps are

Jupiter's giant eye lives on the border- land between order and chaos. This picture was taken by *Voyager 1*.

clusters that contain gaps—just the kind of random yet strangely orderly grouping that is the signature of a fractal. Jupiter's cyclopian eye is also fractal, composed of swirls within swirls, an organized dynamical system woven out of chaos.

In the old scientific aesthetic, the beauty of outer space lay in our perception of an underlying mechanical order. In the new aesthetic, scientists appreciate the universe as a fluctuating holistic hybrid of symmetry and chaos.

Scientists suspect that the gaps in the rings of Saturn may have something to do with chaos. These gaps seem to result from the feedback effects of gravity exerted by Saturn and its satellites, which conspire to make the regions so erratic as to be unoccupiable for any long period of time. In the asteroid belt between Mars and Jupiter lie several orbital gaps, or empty regions, which have almost certainly been created by chaos.

OUR

WEATHER

TODAY

IS

CHAOS

•

It may happen that slight differences in the

initial conditions produce very great

differences in the final phenomena. . . . One-

tenth of a degree more or less at any point,

and the cyclone bursts here and not there,

and spreads its ravages over countries which

it would have spared. This we could have

foreseen if we had known that tenth of a

degree, but the observations were neither

sufficiently close nor sufficiently precise, and

for this reason all seems due to the agency of

chance.

—Henri Poincaré, great nineteenth-

century physicist, perhaps the first

scientist to confront the perplexities of

dynamical chaos.

•

Meteorologist Edward Lorenz's cup of coffee is world famous. When he left the computer at his Massachusetts Institute of Technology lab to go have it, he didn't suspect that the turbulent swirls of steam he saw rising above the cup's rim were emblematic of the revolutionary chaotic message his computer was at that very moment calculating. In the past two decades, magazines and books all over the world have reported what Lorenz found when he went back to his computer and looked.

Lorenz had been working on a simple three-variable model for forecasting the weather, and his computer had cranked numbers into the model's nonlinear equations to make a forecast. He decided he wanted to extend the forecast a few more days, so he needed to make another computer run. Since computers were relatively slow in the 1960s, Lorenz took a shortcut and rounded off some of the numbers he plugged into his model's equations. He expected a slight discrepancy between the two calculations of his forecast, but he was sure it would not be enough to affect what he was looking for. He set the computer to work on the shortcut version of his forecast and went out for his coffee.

Though each emerges from a different set of meteorological conditions, hurricanes and tornados are self-organized forms born of the underlying chaos of weather. They are like the swirling, strangely ordered shapes that inhabit the boundary regions of the Mandelbrot set.

The way the lightning branches and forks creates a fractal pattern. Fractal geometers calculate jaggedness or brokenness of irregularly shaped lines such as lightning to arrive at the line's "fractal dimension." Lines that have detail on many scales are said to have a fractal dimension that lies between the one-dimensional Euclidian line and the two-dimensional Euclidian plane. The fractal line of many lightning strikes is about 1.3, fractal geometers calculate.

When Lorenz came back he discovered chaos.

Lorenz's computer showed him that the small difference in the starting point between the two runs of the forecast had "blown up," as chaologists now put it, leaving him with two very different long-term forecasts. We can see the same effect he noticed in this plot of two long-term forecasts of westerly winds.

The starting information for each of the two long-term predictions of wind speed is very close. The blue plot starts with the wind at 12.00 meters per second and the red with the wind at 11.98 meters per second. For the first fifteen days or so the slight difference leads to very similar forecasts. But then the forecasts diverge radically. Lorenz realized this divergence meant that every forecast is incredibly sensitive to the initial information the meteorologist puts into it. Small errors in that information will quickly balloon to become large errors in the prediction. Any information the meteorologist is missing and does not plug into his model (for example, the information left out when numbers with several decimal places are rounded off by a computer) will end up overwhelming the validity of the forecast.

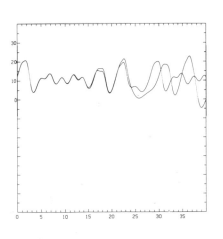

But there was more. Lorenz realized that the problem wasn't just that forecasting models always display limited precision. The problem is that no model, no matter how sophisticated, could ever obtain sufficiently accurate information

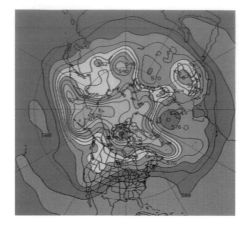

The weather over the North Pole as it looked on May 28, 1991.

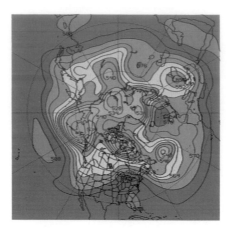

Here is a forecast of the weather for May 30 (two days later) based on the initial conditions in map **A**.

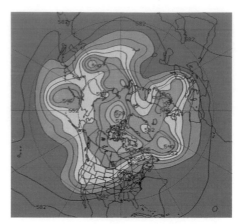

A forecast for June 12 (15 days since the first day when actual measurements were taken) based on the initial conditions of **A**.

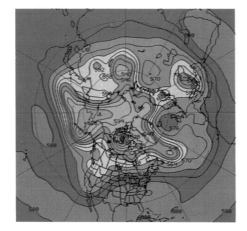

Six hours later on May 28, the weather hadn't changed much. The differences in the initial conditions of the two weather maps are minor. However, projections into the future of these slightly different initial weather conditions reveal the difficulty involved in making long-range forecasts.

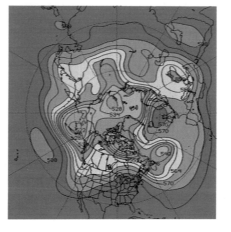

This is a forecast of the weather for May 30 based on the initial conditions in map **B**. Notice how the **A** and **B** forecasts are beginning to diverge at this point.

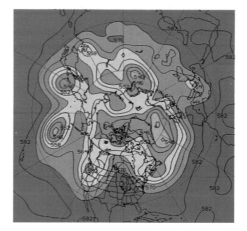

A forecast for June 12 based on the initial conditions of **B**. The **A** and **B** forecasts of what the weather will look like 15 days after the start of the process have now diverged so much that they predict vastly different weather.

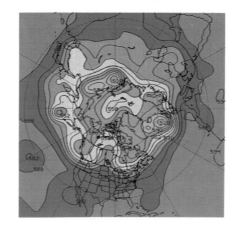

This is what the weather actually looked like on June 12. Because the weather is so dynamical and sensitive to its smallest elements, the accuracy of long-range forecasting will always be severely limited, chaologists say.

to start with because weather itself is so dynamical, so sensitive to the "information" continuously circling inside it, that even the wings of a butterfly stirring in the forests of Brazil would have to be taken into account. Accurate long-term weather prediction, Lorenz concluded, is therefore not just practically, but also theoretically impossible. Aksel Wiin-Nelson, director of the European Centre for Medium Range Weather, put it this way: "We thought that if you just knew the state of the atmosphere sufficiently well and if you built the right models with powerful enough computers, there should be no limit in predicting the weather. Lorenz's work came as quite a shock."

In one sense Lorenz's discovery revealed the obvious. Everything in the vast dynamical system we call weather is connected by feedback to everything else. As a result, just where any "part" of such a system starts out will have an enormous impact on where it ends up. Two specks of ice drifting in the upper atmosphere might start out at almost exactly the same place, but the microscopic differences in each speck's initial conditions will lead each to a vastly different fate. The complex and subtle dynamical forces acting on each individual snowflake as its crystal grows will result in very dissimilar final forms. The exception proves the rule. Look at these two snow crystals discovered by Nancy Knight of NASA's International Satellite Cloud Climatology Project. They are, as Knight says, "if not identical, certainly very much alike." She speculates they fell hooked together as Siamese twins, so the dynamical forces acting on them were virtually the same. Still, we can see there are differences.

Because all of the weather's components (temperature, air pressure, moisture, etc.) are subject to a sensitive dependence on initial conditions—right down to the position and condition of individual molecules in the atmosphere—long-range forecasts will always diverge from actual weather within a few days no matter how sophisticated the forecaster's information.

Lorenz figured out a way to plot the unfolding divergence that took place in his weather model's equations on a graph. The result was a masklike shape called a *strange attractor*. The particular strange attractor that came to be named after Lorenz is an abstract portrait of the infinite raveling and unraveling of the weather as a dynamical system.

The self-similarity and ceaseless change at all scales of activity shown in the Lorenz attractor mean the plot is a fractal.

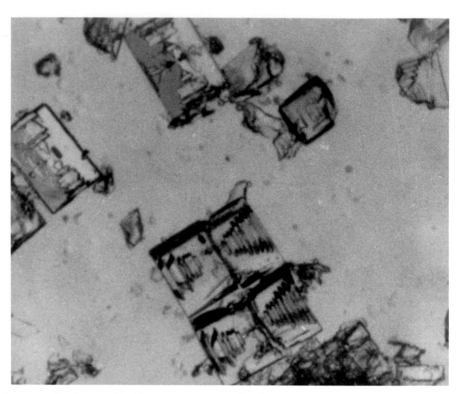

The weather is a quintessential chaotic system. Because of its constant folding back on itself—its "iteration"—it displays a huge range of behavior on many scales, yet remains within the broad limits of behavior we call climate. Climate is another name for the weather's strange attractor.

While they can't hope to ever make perfect predictions, meteorologists are now using chaos theory to evaluate the reliability of their models and to assess whether some initial conditions are more unstable than others. We may soon see a "confidence factor" assigned to our weatherman's five-day forecasts.

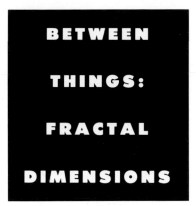

BETWEEN THINGS: FRACTAL DIMENSIONS

•

I coined fractal from the Latin adjective fractus. *The corresponding Latin verb* frangere *means "to break": to create irregular fragments. It is therefore sensible—and how appropriate for our needs!—that, in addition to "fragmented" (as in fractional or refraction),* fractus *should also mean "irregular," both meanings being preserved in* fragment.

—Benoit Mandelbrot.

•

Polish-born, French-educated, American mathematician Benoit Mandelbrot caused a stir in 1967 when he proved that the coastline of England is infinitely long. It was one of several strange conclusions wrought by the new geometry Mandelbrot invented.

The word geometry means "to measure the land." Euclidian geometry measures land by distances, i.e., by angles and lengths. It portrays the land abstractly as blank, smooth parcels composed of points, straight lines, circles, rectangles, triangles, cubes, and spheres. Mandelbrot's revolution was to reveal what everybody knows—that the actual landscape is not smooth or blank at all and that distance is relative. In the real land, space is filled, twisted, kinked, and pocked.

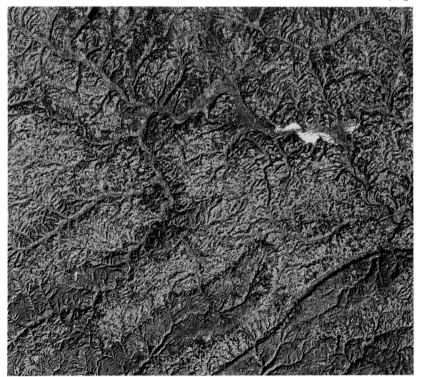

Use the mileage rule at the bottom of a map to calculate the miles between one town and another, and you will probably get fooled. A country road looks straight on the map, but a real road curves and winds across hills. Fifteen miles "as the crow flies" (though it's a fallacy that crows fly straight) will prove more than that when you drive. Mandelbrot showed that distance on the land is relative to scale and detail.

Bend a thread around the coastline of Great Britain on an atlas map and ascertain the length against the map

The two-dimensional surface of this landscape near Elmira, New York, crumples fractally toward three-dimensional space. The city of Elmira is the purple-colored area at the top and to the west of the low-lying white cloud bank trapped in the Chemung River valley. Notes Mark Eustis of the Earth Observation Satellite Company, which produced this image: "The forests of the central Pennsylvania mountains are a russet-green color. They are surrounded by the random-ordered pattern of fields, which are laid out across the hills and valleys of the southern tier in a pattern which seemingly ignores the considerations of runoff and geographic boundaries. This area is a classic illustration of dendritic drainage." Fractal geometry describes objects and processes that inhabit the countless niches between our conventional Euclidian dimensions. The fractal pattern of this landscape was produced by a confluence of chaotic forces.

scale. Now do the same with a more detailed nautical map of the U.K. coast. Oddly, on the second map the coastline is longer. If you were to take a pair of stout boots, a meter-long rod, a few months vacation, and actually measure the British coastline on the ground, you'd find it had grown once more. Use a ruler

Magnify a small portion of a coastline and it looks similar to a larger-scale version. Draw a line that follows the convoluted edge of a coast, and that line's complexity will fill a space between dimensions. A coastline is a fractal pattern left behind by the forces of dynamic chaos.

a centimeter in length and it will be vastly longer still. The reason for this protean expansion of the British coastline is that the smaller the ruler you use, the more of the coast's twists and gnarls you can measure. Imagine how long your result would be if you could measure the molecules along the water's edge with a ruler a scant photon in length. Mandelbrot argued that in order to appreciate how the points, lines, planes, and solids of the real world fill space, the Euclidian idea of distance (and measure) must be abandoned.

Mandelbrot put this idea together with some insights he had gleaned from several mathematicians whose work at the end of the nineteenth century had challenged the Euclidian concept of dimension. These mathematicians, who included the German Karl Weierstrass, the Italian Giuseppe Peano, and the German Helge von Koch, had shocked their colleagues by creating curved lines, so-called monster curves, that convoluted in such intricate ways they could entirely cover the surface of a plane. The result was a disturbing ambiguity about whether a monster curve was a line of one dimension or a plane with two dimensions. Many other monster curves have since been created. Here's one called a "Hilbert curve" (to clarify: mathematicians dub any line that has bends in it a "curve"). The Hilbert curve is generated by starting with a simple figure:

Next, that same figure is applied to each of its own three sides, and some erasing is done. The result looks like this:

Now iterate (repeat) the figure several more times, applying it to itself in the same way as above, and watch as space begins to fill.

Theoretically, the iteration of the Hilbert curve can be carried out indefinitely, so that the curve crosses every point on a plane without crossing itself—hence the ambiguity. Is the resulting figure still a single, one-dimensional line or has it become a two-dimensional plane?

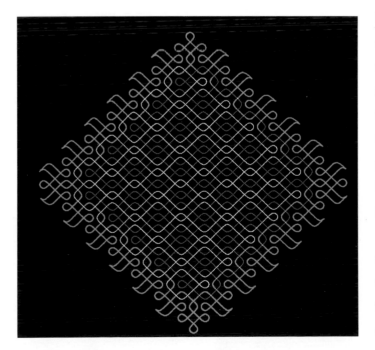

Natives of India who have been taught how to draw a Kolam, as this pattern is called, can draw one very fast, according to biomathematician Przemyslaw Prusinkiewicz. Prusinkiewicz generated this "Soissors" Kolam on a computer using an iterative algorithm similar to the algorithm that produces the space-filling Hilbert curve. He says he finds it "amazing that people in India for centuries were using a fractal as a form of art."

There are many ways fractals are important to art. For example, architects have recently realized that the irregular surfaces of concert halls fractalize the smooth sine waves of the orchestral instruments and enrich their sound.

Nineteenth-century mathematics struggled with monstrous space-filling curves like this one as well as with eaten-out lines like the "Cantor set," and decided it was best to inject them with a theoretical formaldehyde, metaphorically pack them in jars, and store them on a back shelf marked "curiosities"— odd anomalies having no relevance to the rational progress of geometrical knowledge. Then in the 1960s Mandelbrot took them out, dusted them off, and examined them in the light. He saw these pathological shapes between dimensions—"fractals" he called them—as an important clue to a new mathematics of natural forms such as clouds, trees, and mountain ranges.

The classical fractals—those nineteenth-century griffins—are made by adding or taking away elements in a recursive or iterative process. We saw that process at work in the generation of the Hilbert curve. Here's the generation sequence of another classical fractal called a Koch island—made by repeatedly adding a triangle to the middle of every straight line at each iteration. The first generation is at the bottom.

Though the repeated triangle shapes make this fractal perfectly symmetrical, Mandelbrot realized that the Koch island does suggest the kind of detailed, recursive jaggedness that exists along a real coastline.

He discovered that classical fractals had also been made by repeatedly (iteratively) taking something away. The simplest example of this type of iteration involves subtracting the middle third of a line and then repeating that operation indefinitely to create a "dust" of points, called Cantor dust after its discoverer, the Russian-born, German mathematician Georg Cantor.

These ancient bubbles trapped in sandstone show a natural fractal scaling reminiscent of Cantor dust. The dynamic chaos of bubbling has left behind this fractal pattern.

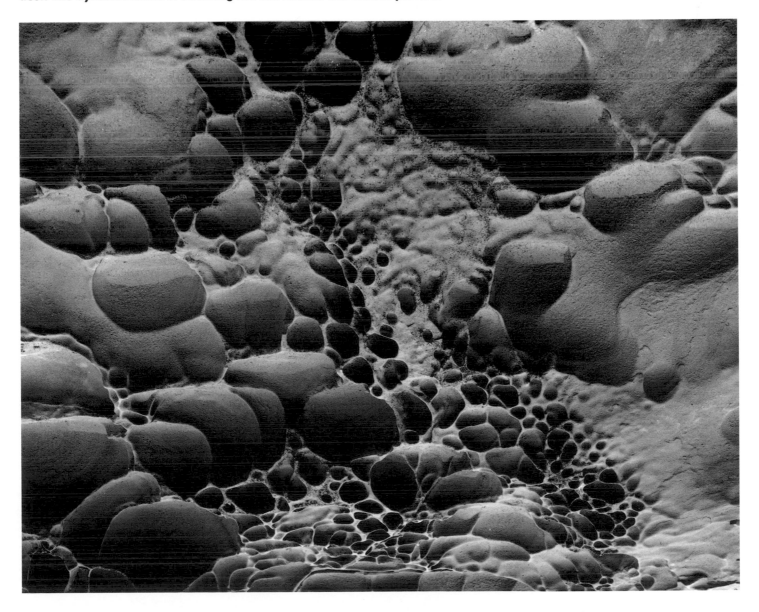

Like the Koch island, this dust is suggestive of structures in nature—though, again, it is too symmetrical—for example, of the way stars are distributed in clusters and dusts across the night sky.

A more complex form of taking something away to create a fractal is an iteration where triangles are repeatedly subtracted from within triangles, creating a figure called the Sierpinski gasket.

The end effect of these subtracting iterations is to shrink the two-dimensional triangle into a figure that fills the space between one dimension and two.

The Sierpinski iteration can be carried out on a three-dimensional object, by subtracting pyramids from within pyramids. The result is called the Sierpinski arrowhead. The example here was fancifully generated on a computer as a "Desktop Tetrahedron" by scientists at the University of Regina in Canada. The arrowhead has more spaces than a sponge and lies about halfway between a two-dimensional surface and a three-dimensional pyramid.

The classical fractals Mandelbrot first contemplated are now called "linear fractals," a name indicating that as the iterations proceed, the lines in the figures stay straight. Put another way, this means that the feedback loop of iteration that generates the figure is well behaved and regulated, so that the figures produced are exactly self-similar on many scales. Magnify a small portion of the Sierpinski arrowhead, then magnify a small portion of the first magnification. The two magnifications will look exactly the same. Compare this with what happens when you magnify a portion of a classical Euclidian figure such as a circle at increasing powers. As your examining lens gets stronger, you will see a smaller segment of the circle's curve, and the curve will look increasingly like a straight line. No new detail is revealed as magnification increases. Magnify parts of fractals, however, and they reveal new, though self-similar, detail.

Self-similarity and scaling are characteristic of fractals in general, Mandelbrot realized, though, not all fractals are as symmetrical in their scaling detail as are the linear, "classical" fractals like the Sierpinski gasket and the Koch island. Mandelbrot discovered that by using what are called "nonlinear" equations, the feedback of iteration that produces a fractal can bend straight lines into curves and swirls and make the self-similarity at different scales variously deformed and unpredictable—a "statistical self-similarity." the Mandelbrot set is perhaps the best-known example of a nonlinear fractal, though it exists in a purely mathematical realm. The swirls and Roman candle explosions that appear along the edge of this mathematical object create a coastline of infinite self-similar intricacy. (See *Mandelbrot Set.*)

Eventually a third type of fractal was discovered that introduces a random element to the iteration. For instance, by randomly changing the size and shape of the triangles as they are iterated inside of triangles, the irregularity of a mountain range can be imitated. The random fractal allows fractal artists to model the natural roughness and irregularity of surfaces such as waves, clouds,

The ragged, expanding boundaries of forest fires are as fractal as a coastline.

In an orchard, disease also progresses like a fire, and scientists have used their knowledge of fractals to figure out how many trees should be randomly excluded from an orchard's rows to prevent disease from spreading.

mountains, and the branching patterns of trees. (See *Imitations.*)

But whether the fractal is classical (linear), nonlinear, or random, the complex way it fills space establishes it as an object between dimensions. Thanks to Mandelbrot's geometrical invention, mathematicians and scientists can now calculate the fractal dimensions of virtually any wrinkly, crinkly, or dendritic object which has details at many scales—from mathematical objects like the Mandelbrot set to natural objects like trees, to human-made objects like Swiss cheese.

Simply put, the fractal dimension indicates the degree of detail or crinkliness in the object, how much it occupies the space between the Euclidian dimensions. The rugged coastline of Britain is a line crumpled up enough to partially fill a plane. Using techniques that Mandelbrot developed, scientists now describe this coastline as a fractal with a dimension of 1.25, similar to the Koch island curve, which has a dimension of 1.2618... —in other words, about a quarter of the way between a line and a plane. The Sierpinski gasket is a fractal with a dimension of 1.584...; protein surfaces bump up and wrinkle around toward three-dimensional space in a dimension that is around 2.4. In a fractal world some objects have been found to be incredibly complicated in the way they fill space. Mathematicians have recently proven that the edge of the Mandelbrot set is so intricate that it is a one-dimensional line with a fractal dimension of

The cauliflower—a fractal in your refrigerator—arranges its florets in self-similar scales. Self-similarity, in this case, is a pattern left behind by the dynamical process of growth as it filled the space between dimensions.

Imagine a piece of paper as a plane of two dimensions. Now wad it up. The resulting object is neither a plane nor a sphere, but something folded in between the second and third dimensions. As calculated by fractal geometry, this wad of colored paper has a fractal dimension of about 2.5.

2. The two-dimensional surface area of the human vascular system is folded, bent, and packed so extensively that it has an effective fractal dimension of 3; the system of arteries alone has a dimension of 2.7.

Most natural objects, including ourselves, are composed of many different types of fractals woven into each other, each with "parts" that have different fractal dimensions. For example, the bronchial tubes in the human lung have one fractal dimension for the first seven generations of branching, and a different dimension for the branching after that. In the complex environment of nature, intricate patterns of self-similar, scaled detail were laid down by the dynamical forces affecting evolution, growth, and function.

When he first published his ground-breaking book, *The Fractal Geometry of Nature*, in 1977, Mandelbrot defined the concept of fractals in terms of the mathematical methodology used to calculate the fractal dimension of any object or process. In a later edition he regretted having proposed a strict definition of fractals at all. He writes that "for me, the most important instrument of thought is the eye. It sees similarities before a formula has been created to identify them." We will recognize fractal patterns intuitively long before we specify them logically and mathematically. Leaving a definition open is not an unusual procedure in math and science, and seems especially fitting for the idea of the fractal. Not only does such openness allow us to explore the richness of the concept without arbitrary restrictions, it emphasizes the great shift fractal geometry has made away from a strict quantification of nature—measuring objects and processes in terms of degrees, lengths, and calibrated time durations—and toward an appreciation of the *qualities* of nature such as roughness, openness, branchiness, and roller-coaster rides of "fractal time." When we are

not confined to a strict definition of fractal geometry, we can appreciate that this geometry is more than a measure of nature; it is a way of concentrating our attention on the rich activity that has long been taking place in the vast, busy spaces and cracks overlooked by our old quantitative Euclidian perception.

THE

HAUNTING

MANDELBROT

SET

•

Zoom in at any part of the set

at any magnification, it

always reveals a reproduction

of itself. As the zoom

continues, the same image

reappears ad infinitum. In The

Poetics of Space, *Gaston*

Bachelard [says] . . . the

scientist "has already seen

what he observes in the

microscope and,

paradoxically, one might say

that he never sees anything

for the first time."

—KLAUS OTTMANN, art

curator and

contributing editor

of *Flash Art.*

•

Largely because of its haunting beauty, the Mandelbrot set has become the most famous object in modern mathematics. It is also the breeding ground for the world's most famous fractals. Since 1980, the set has provided an inspiration for artists, a source of wonder for schoolchildren, and a fertile testing ground for the science of nonlinear dynamics. It is the symbol of the Chaos Revolution.

The set itself is a mathematical artifact—an odd-shaped infinite swarm of points clustered on what is known as the "complex number plane." Let's try to visualize it.

IBM scientist Clifford Pickover employs some clever tricks to bring out a different kind of detail in the Mandelbrot set frontier. He calls this image "Mandelbrot Stalks."

To make them tangible, we imagine real numbers like 1, 2, 3 . . . as spaced out along a number line. Because complex numbers have two parts to them—called their "real" and "imaginary" parts—making complex numbers tangible requires two lines, or axes, which means a plane. Picture the plane dotted by complex numbers as a computer screen, which is just where the visual form of the Mandelbrot set was discovered. Like the screen of your television set, a computer screen is covered with a host of very tiny, evenly spaced points, called pixels. The moving image on the screen is made when patterns of pixels are excited (made to glow) by a fast-moving scanning beam of electrons. Think of each pixel as a complex number. The pixels in any neighborhood are numerically close to each other, just as 3 and 4 are numerically close to each other on the real number line. Pixels (numbers) are made to glow by applying an iterative equation to them.

In the late 1970s and early 1980s Benoit Mandelbrot, the inventor of fractal geometry, and several others were using simple iterative equations to explore the behavior of numbers on the complex plane. A very simple way to view the operation of an iterative equation is as follows:

Start with one of the numbers on the complex plane and put its value in the "Fixed Number" slot of the equation. In the "Changing Number" slot put zero. Now calculate the equation, take the "Result" and slip it into the "Changing Number" slot. Repeat the whole operation 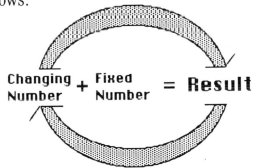 again (in other words, recalculate or "iterate" the equation) and watch what happens to the "Result." Does it hover around a fixed value, does it spiral toward infinity quickly, or does it stagger upward by a slower expansion?

When iterative equations are applied to points in a certain region of the complex plane, the results are spectacular. By treating the pixels on computer screens as points on the plane, even nonmathematicians can now admire this marvel. In fact, without computers, only the most intuitive mathematicians could have glimpsed what was there. With the computer it works like this:

Starting with the value of a point (or pixel) and applying the equation to it, iterate the equation perhaps 1,000 times. If the "Result" remains stable, color

This flamelike image of the set has been named "Peitgen" by Homer Smith after the German mathematician Heinz-Otto Peitgen, who brought the beauty of fractals forcefully before the public eye.

the pixel black. If the number heads at one speed or another toward infinity, paint it a different color, assigning colors for each rate of movement. The points (pixels) representing the fastest-expanding numbers might be colored red, slightly slower ones magenta, very slow ones blue—whatever color scheme the fractal explorer decides. Now move on to the next pixel and do the same thing with the color palette until all the pixels on the screen have been colored. When all the pixels (or points representing complex numbers) have been iterated by the equation, a pattern emerges. The pattern that Mandelbrot and others discovered in one region of the complex plane was a long-proboscidean insect shape of stable points—the Mandelbrot set itself, usually shown in black—surrounded by a flaming boundary of filigreed detail that includes miniature, slightly distorted replicas of the insect shape, and layer upon layer of self-similar forms.

The boundary area of the set is infinitely complex, therefore fractal, because it is possible to bring out finer and finer detail. Computer graphics artists call the process of unfolding the detail "zooming in" on the set's boundary or "magnifying" it. It's fairly easy to grasp what this means.

On the real-number line we routinely imagine that between the numbers 1 and 2 are other numbers, 1.5, for example, or 1.6. (We encounter this every time we pick up a ruler.) Of course, between those numbers are still more numbers, 1.53 and 1.54, for example—and so on, indefinitely. The same is true for the numbers on the complex plane. Between any two of them are many more, and between those many more are many more still ad infinitum. These numbers between numbers allow us to use the computer like a microscope to

By iterating the points between the points at one scale of the Mandelbrot set, it is possible to zoom into increasingly smaller scales of the set. Because there is an infinite number of points between any two points, the Mandelbrot set's detail is infinite; a coastline complex beyond measure. The first frame here shows the region of the number plane where the set resides—with the set itself in black and the fractal boundary area on fire with color. After that, each of these 12 frames explores increasingly deeper magnifications of detail in the fractal boundary.

continued on next page

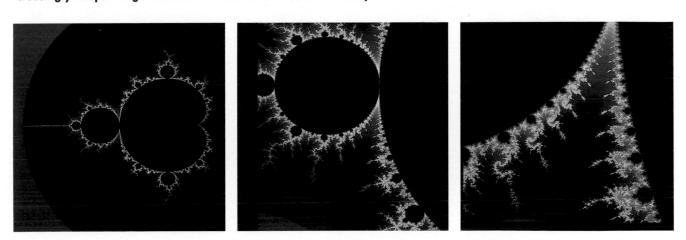

dive into increasingly deeper detail. To extend our analogy, if the numbers we were examining on the complex plane were all like the numbers at the level of, say 1, 2, 3, etc., on a ruler, then we would be examining the largest scale of numbers. But we could also go to a smaller scale and examine the numbers at the level of 1.5, 1.6. Between those will be yet a smaller scale (including the numbers 1.53 and 1.54, for example)—and so in any region of the complex plane we could move downward (or inward) to smaller and smaller scales.

Similarly, explorers of the Mandelbrot set can zoom in to study finer and finer detail as they examine the ever smaller scales of numbers between numbers on the complex plane. Indeed, a home computer can examine numbers out to 15 decimal points. To complete the microscope analogy, if the numbers 1 and 2 were the equivalent of objects the size of human beings and trees, a number 15 decimal points smaller would be an object tinier than an atom. More powerful computers can go into even finer (or deeper) detail. In addition, different styles of iterative equations can act as prisms to display varying facets of the behavior of the complex numbers around the set.

Applying zoom-ins and different iterative prisms to the numbers in the boundary area of the Mandelbrot set has revealed that this region is a mathematical strange attractor. The "strange attractor" name here applies to the set because it is self-similar at many scales, is infinitely detailed, and attracts points (numbers) to certain recurrent behavior. Scientists study the set for insights into the nonlinear (chaotic) dynamics of real systems. For example, the wildly different behavior exhibited when two numbers with almost the same starting value and lying next to each other in the set's boundary are iterated is similar

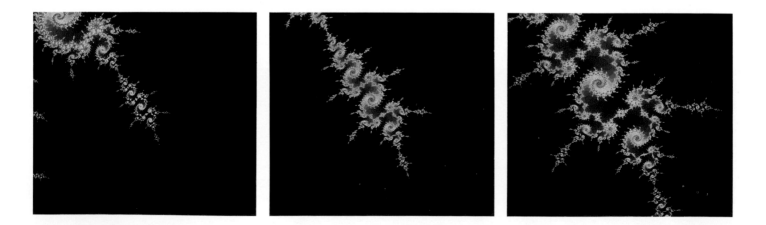

This jewel box version of the detail on the Mandelbrot set was computed by Rollo Silver of Amygdala in San Cristobal, New Mexico. Silver also puts out a newsletter aimed at fractal fanatics.

to the behavior of systems like the weather undergoing dynamic flux because of its "sensitive dependence on initial conditions."

But a major importance of the set may be that it has become a strange attractor for scientists, artists, and the public, though each may be drawn to it

continued on next page

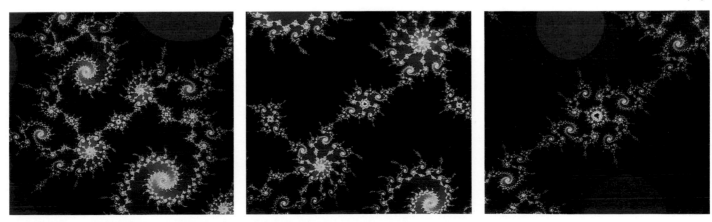

for quite different reasons. Scientists have found themselves attracted—often with childlike delight—to a new aesthetic that involves the artistic choices of color and detail they must make when exploring the set. Artists and the public have been attracted by the set's haunting beauty and the idea of abstract mathematics turned into tangible pleasures.

As testimony to the popularity of the set, Art Matrix, an independent research group based at Cornell University in Ithaca, New York, has sold half a million postcards and countless videos of Mandelbrot fractals since it started in 1983. Founded by Homer Smith and Jane Staller, Art Matrix grew out of Smith's work with Cornell mathematician John Hubbard to produce images for Hubbard's research. A deep thinker, Hubbard had proved one of the important theorems about the set—a holistic theorem, that all the mini-Mandelbrot figures folded into the boundary are mathematically connected. He was also one of the researchers who decided to name the set after Mandelbrot in recognition of the French mathematician's role in bringing its outline to light. Smith reports that a collaboration among Benoit Mandelbrot, John Hubbard, and Heinz-Otto Peitgen led to the *Scientific American* August 1985 cover article, and thousands of requests from readers for views of the set to hang on their walls. This collaboration also led to a friendly rivalry over who could produce the most aesthetically pleasing renditions of the set. Smith continues to help Hubbard with his research and to produce images for the public with the aim of attracting young children to mathematics. "We hope that fractals show up in early classrooms, to get kids interested in mathematics very early," Smith says, "because it really

This image, which Homer Smith of Art Matrix calls "The Orchid," is a part of the Mandelbrot set explored by the iterative equations of "Newton's method," a mathematical technique used to solve polynomial equations. (See *Math Art.*)

opens the eyes of children who haven't been turned off to education. . . . We hope by the time they get up to the tenth grade, they'll have seen these things and say, 'There's something here in math, science and computers that I want to learn.' "

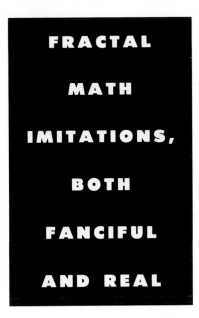

FRACTAL MATH IMITATIONS, BOTH FANCIFUL AND REAL

•

Scientists will . . . be surprised and delighted to find that now a few shapes they had to call grainy, hydralike, in between, pimply, pocky, ramified, sea-weedy, strange, tangled, tortuous, wiggly, whispy, wrinkled, and the like, can henceforth be approached in rigorous and vigorous fashion.

—Benoit Mandelbrot, inventor of fractal geometry.

•

Without knowing it, millions of people around the world have watched fractal mathematics on movie and television screens. Using variations of fractal techniques originally pioneered by Benoit Mandelbrot and IBM researchers, computer graphics artists created the alien landscapes for the *Star Wars* films and *Star Trek II: The Wrath of Khan*. Fractals have become an important staple of Hollywood special effects.

Lucasfilm production used fractals in their epic *Star Wars* films. The program that made this image generated the mountains in the "Genesis Demo" segment of *Star Trek II: The Wrath of Khan*. This is an early mountainscape made by Loren Carpenter of the film company's computer graphics division (now a separate company known as Pixar), using the midpoint displacement method. Back in the

early 1980s Carpenter had seen some of the fractal images made by Mandelbrot's colleagues at IBM. "I saw the picture of the mountain range and said, 'Hey, I've got to do this!' But the methods Mandelbrot uses are totally unsuitable for animation, for making a picture where you stand in the landscape." So Carpenter developed his own methods for fractal animation and landed a job on the *Star Wars* projects. With computer animation, he says, "You can recreate the pyramids or a civilization from another planet. You can change colors, twist or deform shapes, do things that are completely fantastic. Fractals are an excellent tool to extend the range of possibilities."

Researchers have learned that relatively simple mathematical formulas can be used to model the self-similar patterns in a natural object such as a mountain range. Since the pattern at smaller scales repeats at larger scales, by using formulas that involve recycling numbers again and again—iterating them— patterns can be made to evolve into imitations of large-scale real-world objects.

One of the earliest techniques for generating fractal imitations of mountains, for example, involves the simple, repetitive action of randomly displacing the midpoints of triangles. A fairly simple formula tells the computer to draw a triangle inside of a triangle after randomly moving, or displacing, the midpoints on each of the original triangle's three sides. As the iterations proceed, the shape of each successively smaller triangle is altered, and the expanding jumble of triangles within triangles grows into a mountainscape. Midpoint displacement is illustrated in this rough example of two stages involved in generating the image of a planet rising over lunar terrain.

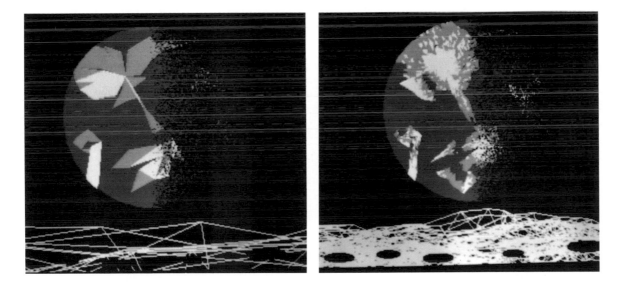

In sophisticated versions of this program, the random amounts that the mid-points are displaced up or down can be adjusted by a "distribution law" which approximates the relative roughness of the real-world terrain being modeled.

Trees and plants can be simulated by recursive programs which contain instructions for drawing repeated shapes to create twigs, stems, leaves, and flowers, while randomly rotating them or bending them, and changing their thickness after a certain number of iterations. By carefully adjusting parameters

and randomness, Przemyslaw Prusinkiewicz of the University of Calgary, Canada, has been able to generate imitations of specific botanical forms, such as the plant *Mycelis muralis.*

Many of the fractal formulas that have been used for simulating botanical forms or landscapes were discovered by trial and error. However, Michael Barnsley of the Georgia Institute of Technology has developed a fairly easy method to find the iterative codes needed to generate even complex scenes. The key, again, is the idea of fractal self-similarity.

Beginning with the object he wants to model, Barnsley shrinks it and distorts it on the computer until he can come up with a series of tiles or "*transforms*" (shrunken and distorted versions of the original object) that can be overlapped and fitted together to create the original large-scale shape. A simple example is Barnsley's fractal model of a maple leaf. "With a picture of a leaf," Barnsley explains, "you've got to say, 'This corner looks like the leaf, if only I squeeze it and distort it and turn it about. This piece is a distortion of the whole thing.' If you make enough of these assertions—even if they're not particularly accurate—then you've essentially written a fractal description of the object."

In this example, there are four transformations of the maple leaf. By keeping track of the stretching and shrinking procedures required to create each transformation (each tile), Barnsley derives a series of *transform* formulas. Then he plays what he calls the "chaos game." Start with a point on the computer screen, mark it, and apply one of the transform formulas. That leads to a second point on the screen. Mark that and apply a transform formula to get a third point, and so on. Which transform formula is applied to a point is determined by a probability that Barnsley calculates and includes in the rules he gives to the computer. All the rules are iterated for a while. As the iterations of the chaos game proceed, the point hops around, leaving apparently random tracks on the screen. However, as the dots pile up, a shape resembling (though not identical to) the original leaf begins to emerge. The original shape has become an attractor for the points as they are shifted from one spot to another by the formulas.

Przemyslaw Prusinkiewicz, a mathematical biologist at the University of Calgary in Saskatchewan, Canada, admits that he derives considerable aesthetic pleasure from the botanical forms he creates on his computer, but he insists that these images are primarily models to help him verify scientific hypotheses about growth. By viewing the growth of plant forms through the lens of fractals, Prusinkiewicz has developed a keen sense of the "deep relationship between self-similarity and growth

continued on next page

rules. I can understand the growth process by seeing what I need to do in order to create a self-similar form on the computer."

He argues that self-similarity is a form of symmetry, and that symmetry is a key concept in modern science. In physics, for example, the notion of matter and antimatter comes from symmetry, and the fact that there is (apparently) more matter in the universe than antimatter indicates that "symmetry-breaking" is one important way that nature creates form. "How can a circular egg develop into a form such as a bird. That is a problem of symmetry-breaking we encounter in biology. The problem of self-similarity fits into this general framework. We are asking, is this self-similarity we see in plants ideal or is it broken to some extent? Looking at a plant, I am asking what kind of growth rules lead to this broken self-similar structure."

Prusinkiewicz notes that the breaking of rigid self-similarity is a long-known principle in art and seems to be a place where art, nature, and science converge. "The artistic content is introduced the moment there is a departure from strict self-similarity," he says. "The material universe also follows this principle. If you look at a real carrot leaf, you see departures from strict self-similarity. When

I try to create a form with fractals, I depart from strict self-similarity to represent the sun attracting a branch, or a branch wilting. The images that result have some artistic value to me, and I cannot create them using another medium. I can represent more than just mechanical things using the program. It can be as emotionally charged as using a paint brush."

Though his work is in fractals, he emphasizes that the living forms he studies are "not at all chaotic." Chaotic systems amplify feedback to transform themselves into new regimes. Living forms use feedback to *manage* change and remain relatively stable. For example, if some cells are removed from the first few iterations of cells in a developing embryo, the organism's feedback will reconfigure the growth process so as to yield a normal form despite the interruption.

To make his computer simulations of botanical forms, Prusinkiewicz uses a sophisticated recursion program. This program not only adds randomly self-similar new growth with each iteration of his formula, it adds die-back and wilting, and the effect of hormone changes on previous growth. The computer imitation of the plant unfolds in a way that is analogous to real growth. "Growth is constant and not sequential. We try to capture the interaction between parts. We could try to reproduce it by capturing every detail, but you would not gain any understanding. We are trying to sort out the key laws and principles, and abstract these from all the things which are irrelevant. This is the essence of scientific process."

But the artistic is always in his mind, too, as indicated by the additions he made to his abstract rendering of the background for his fractal carrot leaf and his water lilies scene, which was specially processed by a color replacement computer program to make it look like an impressionistic painting.

With any of the fractal methods for imitating objects, separate iterative formulas can be used for different parts of the object (for example, one formula for the branching, another for the leaves) or for different elements of a scene. This procedure creates immense possibilities for re-creating the images of complex forms out of relatively simple sets of equations. Fractal technique allows complex information about fractally shaped objects to be stored or "compressed," making efficient use of computer time and memory space—and leaving room for even greater complexity.

Computer graphics artists are not just using fractals to store scenes and create entertaining landscapes. Fractal geometry is being regularly applied to such problems as visualizing how polymers, dentrimers, and other large molecules grow and evolve through random iterations of self-similar dynamical activity.

Traditionally we have used Euclidian shapes—circles, squares, and triangles—to model figures and landscape. It was a process that tended to generalize and idealize the natural world. Fractal geometry brings us perhaps a little closer to nature's infinite subtlety.

Peter Oppenheimer was inspired to take up fractal research by Benoit Mandelbrot himself, who visited Princeton in 1978 where Oppenheimer was studying mathematics as an undergraduate. In his work at the Computer Graphics Lab at the New York Institute of Technology, Oppenheimer perfected using the computer to make fractal imitations of real forms, but has grown skeptical about what they prove.

"Science likes to think its goal is to make objective representations of nature, but it seems to me that all such representations, visualizations, or models merely isolate a few select parameters, a few aspects of the object and say, what happens if we just look at these? Each different approach gives you a slightly different result." He warns, "A lot of knowledge we're gaining from computer pictures is very intuitive and must not be seen to be objective." Science purports to be skeptical of its models, but now it's harder to maintain this skepticism, he thinks, because "our pictures are convincing in a very subtle way. If the picture looks like that object, we figure we must be doing something right." But one of the messages of chaos theory is that no matter how good a scientific model or formula, there is always a fundamental unpredictability and uncertainty driving dynamical systems.

Oppenheimer suggests that contemplating the uncertainty involved in chaos and fractal images may provide a new kind of knowledge. "I don't think we've figured out just what kind of knowledge it is. One reaction to all this is dismay at the limits of our ability to figure things out, but maybe we have to take some sort of leap of faith. Wow, we can't figure it all out, isn't that wonderful? Let's accept these pictures, but let's accept them as something else than the kind of knowledge we're used to. Maybe it becomes art rather than science. It's still knowledge, but a different kind." "Intuitive" is a word Oppenheimer uses repeatedly.

He says that chaos theory changed his perception of the world: "Just seeing how sensitive things are to their initial conditions has changed my notion of our place in the universe and our ability to make things happen. Everything is so interconnected." He notes, for example, that fractals and chaos

force him to acknowledge the interconnectedness every time he attempts to imitate a form: "You've got to take the environment into account. For every fractal you have of a tree, say, the negative space is also a fractal. What forms that shape is a balance between its structure and the environment's structure."

The ability to make forms "like" nature has also transformed his personal outlook on nature. "When I was growing up as a child, I used to believe that the physical objects around us were somehow fundamental and they were there first. That any of the ideas we had about them were outside the physical objects. Now I believe that ideas, mathematical concepts, abstract notions, dreams, spirits are somehow more fundamental, and that these physical objects somehow grow out of that. I've come to that philosophy based on my exploration of computer graphics. What I'm doing there is taking a bunch of numbers and turning them into something that looks organic or natural like a tree, something that emerged from my manipulation of numbers. Now I've become more Platonist. I believe that there are abstract forms and that physical objects are manifestations of those forms. Both the synthetic image on my computer and the tree outside my window are syntheses from something more abstract. But that doesn't mean we can ever make our own syntheses match with the abstraction of nature that produced the tree."

Accordingly, making fractal imitations puts Oppenheimer in touch with a process "similar to, but not the same as" that which makes the real tree. One difference is that he is aware of what's left out, that there is a hole (or whole) that's not in the picture. "The fact that it is different, that something's left out, is what makes it interesting; that's what is most important about it." He says his work has "evolved from science into art; of course, I'd like to blur those distinctions a little bit." In his images he doesn't attempt to make "realistic"-looking forms but tries to present a fiction that is like the fact. This is evident in the stylized look of his "Raspberry, Garden at Kyoto," and in its surreal twin.

This eerie chaos landscape is a frame from John Lewis's animated short film called *Aliens,* made by using fractal geometry. Lewis came to fractal graphics after formally studying art, writing, and psychology at various institutions and taking a graduate degree from MIT's famous Media Lab, "a good place for people who want to be interdisciplinary."

Lewis describes chaos as "a field which studies complexity without explaining it away." He says that until he studied fractals and chaos, he did not see any way of reconciling the subjective experience that humans have of free will with the scientific assumption that everything in the universe is completely determined by its causes. "The fact that chaotic systems are deterministic but unpredictable is sometimes viewed as a solution to the free will/determinism problem. I do not view chaos as a solution to this problem, but I think it shows that the problem is not closed."

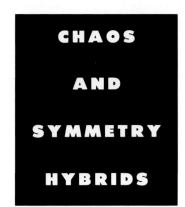

CHAOS AND SYMMETRY HYBRIDS

•

Here is a snowflake in my hand, like some

white Athens in the palm of history,

a moment's fragile Parthenon . . .

. . . And I a god who holds it as it dies

To sudden dew. This molecule of world

May be dominion of a subtler nation,

Inviolate to our eyes. If atoms dream,

What kingdom claims this melting star of

snow!

— ALFRED DORN, from "Snowflake."

•

As a tiny seed crystal falls through the atmosphere, the hexagonal structure of the single ice molecule grows at its unstable boundary by diffusing heat and creating a charge that attracts other water molecules. During the crystal's erratic flight path, its encounter with temperature and humidity affect its pattern as it begins to develop, with one tip or another picking up molecules from the air. A competition between instability at the crystal's boundary and the stability of surface tension across the whole space of its growing mass amplifies the crystal's microscopic preference to grow symmetrically, in six directions at once. Thus the forces of symmetry and chaos combine to branch the crystal's boundary into an intricate lattice form.

Mixing symmetry and chaos is nature's—and art's—common strategy to create form. It is a tension that fires into existence trees, snowflakes, starfish, and our own bodies, and engenders a world that contains both marvelous variety and similarities at many scales.

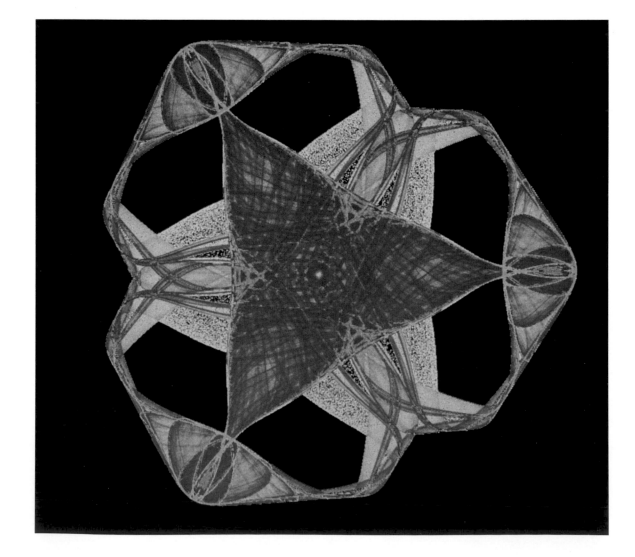

The fractal computer program that produced the snowflakes on the left, like the dynamical forces that produced the real snowflake on the right, combines instructions for sixfold symmetry with the convolutions of feedback that lead to chaos. Notice that the computer flakes seem somewhat unreal because they are too symmetrical. The real snowflake, however, shows that even across the microscopic space of the growing crystal there were subtle differences in the forces affecting

it. Snowflakes are fractal records of the changing circumstances the ice encountered during its descent. No two falling snowflakes will meet precisely the same circumstances. Unique snowflakes demonstrate that the weather is a chaotic system in which all of the "parts" are sensitively dependent on their constantly changing circumstances.

Mathematicians Martin Golubitsky of the University of Houston and Mike Field of Sydney University, Australia, call this bit of symmetrical chaos an "icon." It was generated on a computer using a combination of nonlinear equations which exhibit chaos and equations that involve symmetry. Golubitsky and Field think that in the real world the mathematics they have developed to produce this icon might be used to describe the chaos that occurs in containers such as cylinders, pipelines, and mixing apparatus where the symmetry of the container would affect how chaos unfolds. Golubitsky notes, "Our pictures mixing symmetry with complicated dynamics impose a regularity that was hard to imagine in advance." How many "regular"-looking, even symmetrical, processes and objects in the world have chaos enfolded within them?

(FACING PAGE) This hexagonal pattern of cells formed in a container of heated silicone oil. As soon as the temperature difference between the bottom and the top of the container reaches a critical point, the convection cells bubbling chaotically from the boundaries of the container self-organize themselves so that a symmetry hidden in the chaos asserts itself. Note the beautiful close-up of the self-organized convection cells by Manuel Velarde of the Autonomous University of Madrid.

When the temperature in the container is pushed higher, the symmetrical pattern is maintained and then eventually lost as the system evolves toward a turbulent and chaotic state.

The forces of symmetry and chaos embodied in the DNA molecule both drive life forward and contain it within limits. This is computer simulation of the spiraling DNA ladder as seen from the top.

CHAOS SCULPTS FRACTAL LANDSCAPES

•

We are floating in a medium of vast extent, always drifting uncertainly, blown to and fro; whenever we think we have a fixed point to which we can cling and make fast, it shifts and leaves us behind; if we follow it, it eludes our grasp, slips away, and flees eternally before us. Nothing stands still for us. This is our natural state and yet the state most contrary to our inclinations. We burn with desire to find a firm footing, an ultimate, lasting base on which to build a tower rising up to infinity, but our whole foundation cracks and the earth opens . . ."

—Virginia Woolf, *Pensées.*

•

The physical world we live in is a sea of change, much of it unnoticed by us—or denied. Scientists say that the planet Earth itself is essentially a slow-moving glob of liquid iron surrounded by a slightly faster-flowing glob of liquid rock on which floats a thin crust. On the ocean floor, some of that crust is being sucked into the cauldron beneath, while crustal plates grind into each other, spawning volcano eruptions and earthquakes: fractal and chaotic signs of the immense dynamism of the living place we inhabit.

Since everywhere on Earth's thin crust, the natural landscape is being hewn by chaos into shapes with branches, folds and fractures, and detail inside detail, the immense intermeshing of dynamical forces constitutes the eternal, ever changing dissonance and harmony of nature that has attracted scientists and artists throughout the centuries.

The magnificent power of turbulent chaos shows in this eruption of the Mt. St. Helens volcano in Washington.

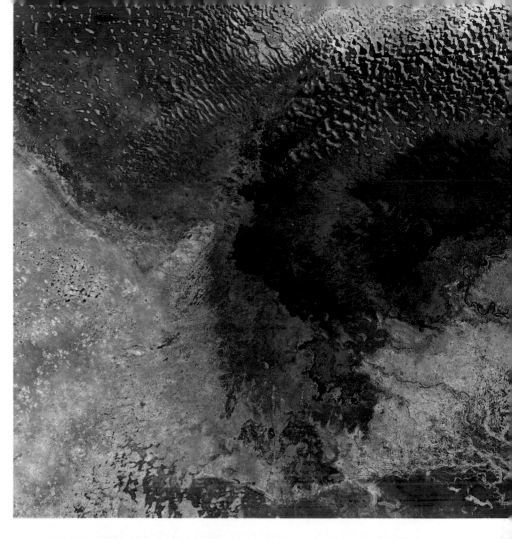

The shoreline and swampy vegetation (in red) around Lake Chad in Africa displays layers of irregular, fractal patterns resulting from the dynamical forces of nature. The greenish areas to the west and northwest are the remnants of the lake prior to a catastrophic drying spell. Mark Eustis, of EOSAT (the company that made this image), says that experience tells him viewers will find this photograph abstract and somewhat difficult to "read." He notes, however, that when scientists have shown members of agrarian or tribal societies satellite images of their homelands, they have been able to read the landscape in them quite quickly, presumably because they are more in tune with the landscape's changes and rhythms.

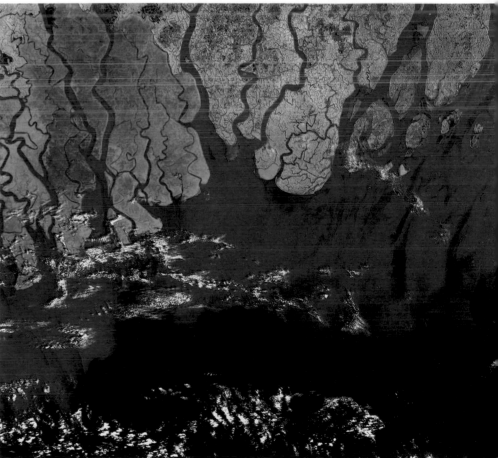

Periodically cyclones—monstrous beasts of self-organized chaos—slam into the mouth of the Ganges in Bangladesh. The tributaries in its delta show the dendritic structure of a classical fractal. Magnify a portion of the drainage system and that section resembles the branching structure of the larger delta. The green area to the west is the last refuge of the Bengal tiger.

Landscape photographer Lawrence Hudetz contrasts the dynamic order of change he finds in nature with the rigid order human beings attempt to impose on nature: "Nature's order is to be continually rediscovered," he says, "That's why it's so exciting, because every photograph is brand new. You're constantly in a different place, a different time. For me, photographs of human

This scene in an ice cave at Washington's Mt. Rainier shows a fractal landscape—the result of many different dynamical processes, including the grinding of glaciers and the stretching, folding, smoothing, and fracturing that results from repeated freezing, melting, and erosion.

Yellowstone Falls demonstrates how water both follows the fractal paths in the landscape and shapes them.

objects don't have this quality of constant rediscovery."

Hudetz's fellow Oregonian, photographer Joseph Cantrell, believes that late-model human beings, driven by the forces of science, technology, and economic self-interest, have worked to trivialize the natural order of chaos. "We dam the rivers, cut the forests, drill the Arctic. It's the attempt to oversimplify, to obliterate the nuance of nature."

Hudetz declares that going out in the field to photograph the fractal shapes of chaos makes him feel "whole and free of inner contradictions," though he knows that may seem a paradox. "Out of all this fractal input I'm getting, I respond to one particular organization at one particular moment in time; when I feel that, the moment speaks to me of this sense of inner freedom." He thinks that his feeling of freedom may come at such an instant because it is the same instant when the photographer recognizes himself, the observer, as self-similar to what he observes.

This portrait of the Oregon woods by photographer Joseph Cantrell might be called "Two Fractals." The dead tree and the rushing stream couldn't be more different, yet Cantrell has captured their deep similarity. The two systems depicted here lie on opposite sides of the dividing line between dynamical order and chaos. The tree was produced by a highly organized dynamical system that resists change; the stream is extremely sensitive and subject to constant fluctuation. From another point of view, however, the chaos of death is overwhelming the tree, while the stream remains a stable, living thing inside its fluctuations. Paradoxes of chaos and order abound in nature.

Cantrell describes photographing nature as a process of sensitizing himself to the subtle movements of nature's creative chaos: "Very early I discovered that if I let things happen as they would happen, I would see something more wonderful than I could create. I think that was an early feeling for fractals. I've never been a good commercial photographer because I don't believe in setting things up. It's almost a form of worship for me to allow things to happen as they will—to be sensitive to the nuance of movement."

Time and weather have eroded the badlands into an ancient kingdom of fractals—a dazzlingly stark ruin of variable self-similarity at many scales.

Photographer Lawrence Hudetz calls this sequence of clouds swirling around Oregon's Mt. Hood "Portrait of a 'Strange Attractor.'" Strange attractor is the name given by chaologists to the plots they chart of the chaos in dynamical systems. Though the plots show that the movement of a chaotic

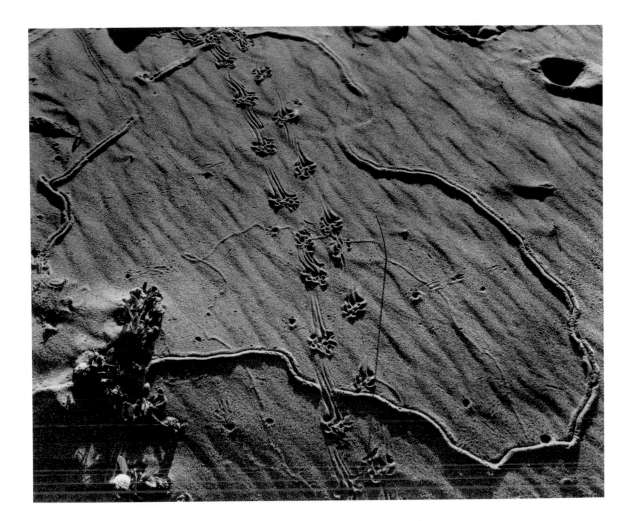

The random movement of organisms on the planet also shapes landscapes. Here a fractal pattern left by the tracks of several creatures was caught by one of America's preeminent photographers, the late Edward Weston. Though biological behavior is underpinned by a form of dynamic chaos making it unpredictable in detail, It clearly displays a subtle, holistic order, here appreciated by Weston.

system never repeats itself and is unpredictable, the system does, curiously, confine itself to a certain region of the plotting space: Chaologists say that the system is strangely attracted to that region, as the clouds in their movement are strangely attracted to Mt. Hood.

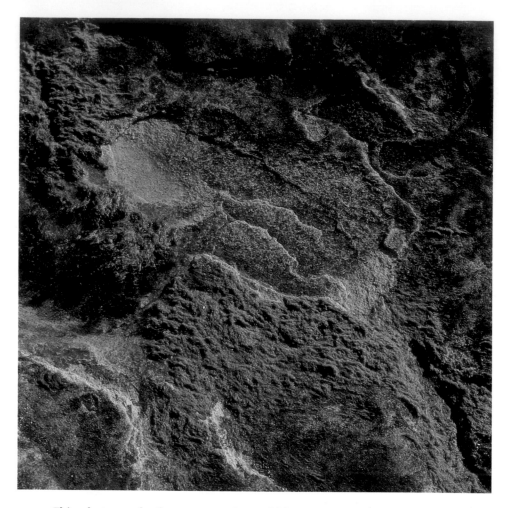

This photograph of moss on rocks could be an aerial shot of trees on rough terrain. Earth has a fractal design because it shows a self-similarity at many different scales.

SPIRALS,

SOLITONS,

AND

SELF·ORGANIZING

CHAOS

•

Chaos-induced complexity is
also partly responsible for our
aesthetic responses. Chaotic
feedback makes, for example,
the amplified guitar playing
of Eric Clapton a more
exciting, complex sound. And
the spontaneous complexity
generated in self-organizing
systems makes a tree more
beautiful than a telephone
pole.

—New Scientist,

October 21, 1989.

•

Nineteenth-century physicists predicted the inevitable drift of the universe toward a heat death of entropy—a random equilibrium without structure. However, in that same century Charles Darwin and Alfred Russel Wallace described a process by which—on Earth at least—more and more complex structure evolves. Could both scientific views be true? The chaologists have largely solved this conundrum.

Thanks to chaos theory and its early pioneers, like Belgian chemist Ilya Prigogine, we now know that *the conditions which give birth to structure are far from equilibrium.* Though in some places (possibly even on average) things may drift toward dissolution, no-thingness, and entropy, in other places there is a natural imbalance—created by chemicals or gases in flux, or thermonuclear energy boiling and spewing into deep space. Out of this imbalance, energized, highly chaotic activity spontaneously produces structure and complexity. The question now being explored is how chaos achieves this magic trick.

These frames show the growth of the Belousov-Zhabotinskii reaction. When scientists plot the equations that describe this reaction, they find that though its activity from moment to moment is unpredictable and chaotic, it stays inside a definite range of behavior. The plot that describes this range of unpredictability is called a Rösseler strange attractor. Some scientists think that a chemical reaction like the BZ gave birth to the first signs of life on earth.

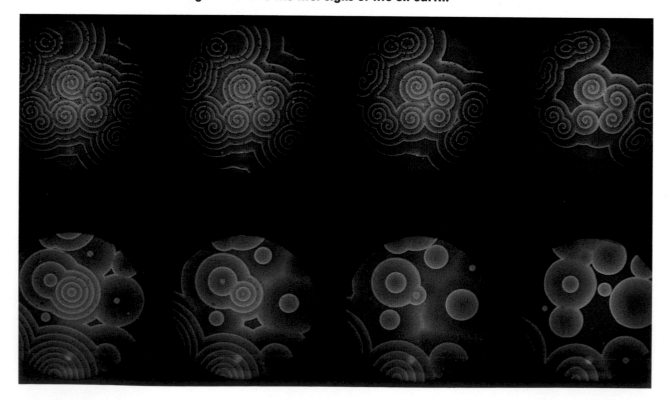

Biomathematician Arthur Winfree, well known for his photographs of the BZ reaction, believes that the spiral lens structure in the compound eye of the firefly may have formed from self-organizing waves. Winfree thinks that the pattern on the lenses is the trace left by an autocatalytic (feedback) process, which brought order from chaos.

Another version of self-organization operates to synchronize the flashes of fireflies when they swarm. On a summer night, fireflies settling in a tree along a riverbank will first flash randomly. Soon, however, small clumps of them begin to flash together, and this synchronization spreads until finally the whole swarm is flashing in unison. Mathematicians studying the phenomenon of pulse-coupled, or phase-locked, oscillators such as electrical oscillators, heart cells, or flashing fireflies have learned something about how phase-locking works. As each oscillator fires, its neighbors are influenced by the feedback of repeated signals so that an oscillator close to its firing threshold senses a signal from its neighbor and fires off immediately. At that point oscillators become locked together. This process proceeds, scientists think, until all of the oscillators (fireflies) become coupled. (For more discussion of self-organization and feedback see the next chapter.)

One of the early clues to the self-organizing process was the discovery of a chemical reaction named after the two Soviet scientists who first described it. The Belousov-Zhabotinskii (BZ) reaction contradicts the long-held belief that chemical reactions are a purely random bonding of reactant molecules. When the chemicals in the BZ reaction are put together in a shallow dish, something curious happens: Characterized by the spontaneous formation of rotating concentric circles, scrolls, and spirals, the reaction looks for all the world like an evolving life-form.

Scientists who have peered into the chemistry of the BZ reaction now know that the order popping up from this chemical fluctuation depends on the formation of a cycle where one of the chemicals begins to produce more of itself, a feedback process chemists call "autocatalysis." The positive feedback of au-

Phase-locked feedback may drive the formation of spiral nebulae—a process that is similar to the Belousov-Zhabotinskii reaction.

tocatalysis acts like a pump creating wave fronts of active regions. Behind these fronts are quiescent regions and adjacent to them are receptive regions into which the reaction proceeds. However, *within* the wave fronts, the same design repeats itself on smaller and smaller scales—making the evolving pattern of the reaction fractal.

When Boris Belousov of the Soviet Ministry of Health first submitted his paper on this chemistry in 1951, it was rejected because his "supposedly discovered discovery was impossible." Belousov never lived to collect the Lenin Prize for his discovery along with Anatol Zhabotinskii of Moscow State University, who verified the reaction, nor to see the chemical journals fill with reports of many other kinds of chemical clocks and autocatalytic reactions. In recent years chaologists have detected the spirals of self-organizing order developing out of chaos in slime mold (whose cell structure at one point in its cycle looks almost identical with the BZ scrolls), in the propagation of signals along nerves, in the formation of spiral nebulae in deep space. Some scientists think that a structure-making chemical reaction like the BZ may have led to the emergence of life on Earth itself.

Paradoxically, scroll-like self-organizing waves can lead to death when they propagate in the electrical impulses of the human heart. Heart attacks and epileptic seizures are, scientists think, a form of self-organized chaos which occurs when the heart or the brain suddenly becomes *too* regular. These body

Scientists have been able to feed the nonlinear equations describing the Be-lousov-Zhabotinskii reaction into a computer and mimic the spiral and scroll-like propagation of these chemical waves. They can also get them by creating "cellular automata." Researchers divide the computer screen into boxes and program simple rules such as "if the boxes to the right and left are empty, grow into the box on the left." A random start, with some boxes filled, leads to a screen filled with chaos. Other random starts flicker and disappear, still others flare into organized forms that propagate across the screen. Researchers have been surprised to find that several quite different sets of rules lead to the scroll-like BZ waves.

This waterspout evolved when a patch of warm air began to rise and caused the surrounding air to eddy and self-organize into a vortex column. Barometric, temperature, and wind conditions must be just right for this ordered structure to emerge out of the flux.

systems lose the variability of their normal, healthy background chaos, and this unhealthy, overly regular state pushes some systems to a critical level, spawning fast periodic waves that pound away at the tissues, causing damage. The "knocking" in your car engine may be another example of an unwanted self-organized oscillation spawned by chaos.

A chaotic system constantly mixes things up, creating new directions in which the system can go. These moments of possibility are called *bifurcation points* by chaologists. At some bifurcation points just the right concentration of a chemical or flux of heat or timing of an electrical impulse can amplify through the system's feedback. The phases or frequencies of the feedback become locked together and a structure emerges.

Once formed, the self-organized structure stays "alive" by drawing nourishment from the surrounding flux and disorder. This is what happens when tornados and other cyclonic winds form out of turbulence. To keep themselves going, they feed off the thunderstorms, moisture, steep temperature and pressure gradients, and turbulence that gave them birth.

Especially long-lasting forms of phase-locked feedback are called *solitons*. Jupiter's eye—first detected in 1644 and apparently enlarged over the next 150 years—is actually a swirling vortex bigger in size than Earth. It was formed out of turbulence, scientists think, at a bifurcation point. There the planet's rotation combined with northerly and southerly layers of high-velocity turbulence to trap a vortex and stabilize it like a piece of dough rolled between the palms of a baker's hands.

The seismic chaos of an earthquake can cause the ocean to phase lock into a "tsunami" or tidal wave a few inches high at the surface but hundreds of meters deep and able to travel intact for many thousands of miles, until it causes havoc by splashing over the continental shelf. Hidden from view, other ocean soliton waves are known to roll vast distances in the boundary between deep cold water and warm water close to the surface.

Technicians have created solitons of light by sending a pulse at just the right frequency down an optical fiber. Unlike other light pulses, the light soliton doesn't disperse over long distances.

As a phenomenon, the soliton was first studied by Scottish engineer John Scott Russell, who came upon a strange wave moving unchanged along a canal and followed it on horseback for several miles. Scientists now know that soliton waves are created when a wave's natural tendency to disperse is exactly compensated for by some critical factor (for instance, the intensity of the light pulse and the size of the optic fiber). Russell's canal soliton kept going because the canal walls and depth created just the right conditions to cause the many wavelets in a turbulent—otherwise dispersing—wave to phase lock. The wave's

A satellite catches several ocean soliton waves following each other. Such waves will travel for very long distances without dispersing. Solitons also have other curious properties. The phases of the elements in a soliton wave are so synchronized that two soliton waves that collide at angles or from opposite directions will pass through each other, emerging on the other side as if no collision whatsoever had taken place.

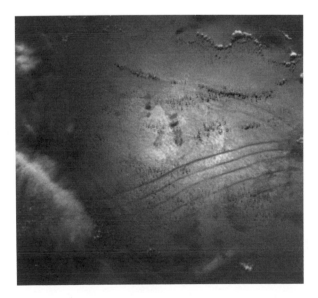

The spiral of life, as this pattern might be called, appears on Stone Age structures around the world. This carving is from Sligo, Ireland, dated at about 2500 B.C. By intuition or some ancient religion-science, the inscribers of these marks seem to have grasped that the spiral pattern symbolizes activity in the life-giving boundary between order and chaos. Anthropologists say the spiral is the ancient symbol for the labyrinth, the twisted pathway for a journey to the core of being.

inherent tendency to disperse was compensated for by the canal walls, which guided the spreading wave back together.

Solitons, like other self-organized structures, breed and thrive in the dynamic world that flourishes on the sharp and delicate edge of dissolution.

Harry Swinney, Joel Sommeria, and Steven Meyers at the University of Texas's Center for Non-Linear Dynamics in Austin created a device that pumps water from an inner ring of six inlets to a middle ring of six outlets so that a rapidly spinning ring of fluid is formed. The pumping action leads to the formation of vortexes which, at a critical rotation speed, merge to form a large stable vortex that mimics Jupiter's Red Spot. This vortex is an eye of stability formed in a crucible of turbulence.

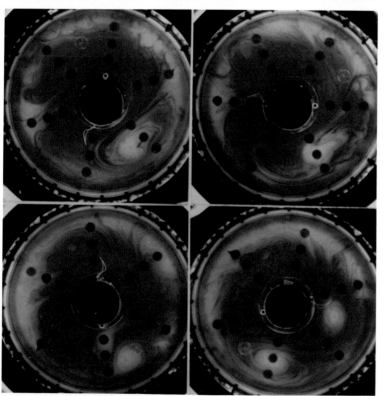

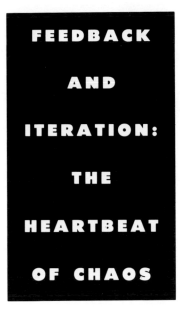

FEEDBACK AND ITERATION: THE HEARTBEAT OF CHAOS

•

For me the peacock's fan has the unmistakable stamp of positive feedback. It is clearly the product of some kind of uncontrolled, unstable explosion that took place in evolutionary time. . . . Darwin compared the [pea]hen to a human breeder directing the course of evolution of domestic animals along the lines of aesthetic whims.

—RICHARD DAWKINS, *The Blind Watchmaker.*

•

You meet a friend you haven't seen in a while. The friend looks different. You say, "Wow, you lost weight," or "You look tired." But you're wrong. You learn to your embarrassment that it's because she has a new hairstyle or he has grown a mustache. Of course. How on earth could you have missed it?

One reason is feedback.

You viewed your friend as a whole, a gestalt, so that every part of your visual image inextricably affected every other part. Change one part and the whole seemed changed. Nonlinear systems—including many dynamical systems and all chaotic systems—are extremely sensitive to small changes, because the feedback among their inextricable "parts" can amplify small changes into large results. A mustache or a new hairstyle isn't much of a change, but the effect on the whole may be impressive.

Scientists usually discriminate between two general, quite different, types of feedback. "Negative feedback" is the type that keeps things in check: The valve on Thomas Watt's steam engine created a negative feedback loop because it opened when the engine was running fast in order to release steam so the machinery wouldn't explode but closed to keep the pressure up when the engine started to slow down. "Positive feedback," which despite its name is not always a good thing, actually pushes systems to explode or spiral out of control. Pointing a TV camera at its own monitor gives the visual equivalent of the positive feedback loop screech that comes from a microphone placed too near its speaker. The frozen frames of the video chaos allow us to notice that there is a structure-making dimension to this positive feedback: This is a place where new forms come into being.

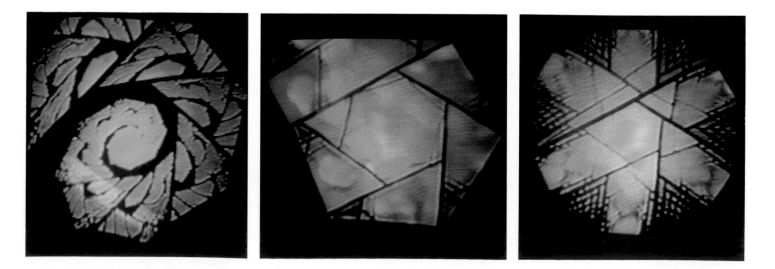

Feedback is everywhere in nature, weaving order out of disorder and holding order tenuously in place. Feedback is the instrument of new life and new havoc—from the positive feedback that escalates an arms race between countries, scrambles computer networks, and sends asteroids flying out of certain orbits, to the negative feedback of pike populations that increase when the trout in a lake get too numerous and fall back when the trout become too scarce.

Richard Dawkins, professor of zoology at Oxford University and author of *The Selfish Gene* and *The Blind Watchmaker,* argues that evolution itself is a grand feedback performance. He notes, for example, that a mutation which improves the design of a predator changes the pressures on its prey so that the prey in turn evolve better defenses in order to avoid these better-adapted predators. As the prey get more wily, the predators once again selectively undergo a design change. Here positive feedback kicks evolution forward. Meanwhile, negative feedback in evolution keeps mutation changes from spiraling out of control— the checking power of many negative feedback loops simply wipes out most mutations and keeps the design of species stable for long periods of time.

Environmental scientists are now debating the role of feedback in the fate of our global climate. On one side are those who believe that the countless

Video chaos results from the iteration that takes place when a TV camera is pointed at its own monitor. In these examples, a mirror was placed at right angles to the screen, and the camera was pointed at the seam where mirror and screen met. The imperfections in the seam were blown up by the positive feedback into chaotic (or fractal) forms made to look like a kaleidoscope by the mirror.

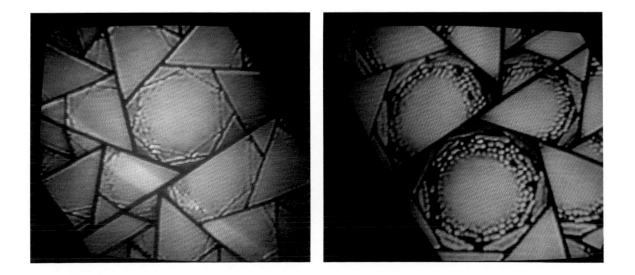

The planet Earth can be seen as a huge dynamical system looped with webs of feedback that keep it relatively stable and evolving at the same time. Positive feedback loops nudge the environment toward change; negative feedback loops keep systems in check. For example, Earth's cloud system acts like the negative feedback of a thermostat to regulate temperature. When the surface of the oceans gets too hot, it gives off water vapor so that clouds form and block out the sun; when the ocean surface cools, the water vapor stops rising, the clouds dissipate and more sunlight comes down to heat up the water again. This photo was taken by the crew of *Apollo 11.*

loops of negative feedback will work to keep the atmospheric temperature stable no matter how much we perturb it by adding carbon dioxide. On the other side are those who point out that spiraling positive feedback somewhere in the system might accelerate even a relatively small perturbation by humans into environmental catastrophe. Because the planet's interlocked positive and negative feedback loops make the global system dynamic—and fundamentally chaotic—it is impossible to predict which fate will befall us.

One of the most important discoveries of chaology has been that positive feedback can cause complex, even chaotic behavior concealed inside orderly systems to unfold, and that negative feedback can grow inside an otherwise chaotic system, suddenly organizing it and making it stable. For example, the chaotic interaction of birds lifting off from the tops of trees exhibits positive feedback. The birds' flight patterns are wild and unorganized as they try to avoid

Feedback develops among the individuals of this school of carp as they try to avoid each other while at the same time being attracted to each other; the carp seem just about to couple through feedback so that they will move in an organized way. Feedback is perhaps the key element in transitions from chaos to order and from order to chaos.

crashing into each other in the first moments of flight. As a result, negative feedback loops are created and suddenly the birds' flight patterns become highly organized. One zoologist has even been able to mimic on his computer the behavior of birds coming into roost by setting up a program with a few simple rules involving feedback such as: birds are attracted to each other but become repelled if they get too close.

Chaologists can mathematically mimic many complex dynamical processes in nature using equations that have terms feeding back from one side of the formula to the other as the equation is run repeatedly, or iterated, on a computer. Iterative equations are now regularly used by chaologists to describe such dynamical processes as the turbulent flow of interstellar gases, static in electrical systems, and the action of reagents in chemical reactions. By using the iterative features of "recursive programing," Dawkins has even created a "biomorph" program that simulates evolution. The program iterates genes and occasionally adds copying errors which are blown up by the feedback into new generations of computer creatures, some of which resemble the *trilobites* that swam in the oceans of the Cambrian era 570 million years ago.

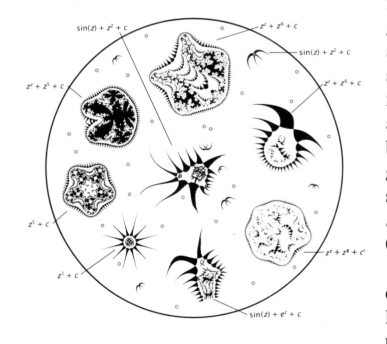

$$\sin(z) + z^2 + c$$
$$z^2 + z^6 + c$$
$$\sin(z) + z^2 + c$$
$$z^2 + z^5 + c$$
$$z^2 + z^6 + c$$
$$z^5 + c$$
$$z^2 + z^6 + c'$$
$$z^3 + c$$
$$\sin(z) + e^z + c$$

Using iterations of purely mathematical functions, Cliff Pickover of IBM created a lab dish of biomorphs.

Pickover's biomorphs show a self-similarity at different scales (small portions of the organism are similar to larger portions of the organism) and illustrate a significant feature of feedback in dynamical systems. Real systems, such as human beings and mountain ranges, also show self-similarity at different scales. The branching of our lungs, nerves, and circulatory systems is evidence that our very bodies are a product of feedback.

Take an equation, solve it; take the result and fold it back into the equation and then solve it again. Keep doing this a million times. That's what Clifford Pickover of IBM did to generate this shape. Each time he solved the equation he marked a point on a graph and therefore he could follow the point as it swept around the plotting space. It's a little like tracing the path of a fly as it whizzes around a room. If indeed this complex shape had been made by a fly, however, it would be a strange fly because it held its flight path to only a certain neatly carved-out portion of the room. The fly would seem to have been irresistibly attracted to that region, though within that region its path was chaotic. Put in other terms, Pickover's feedback sculpture is what scientists call a "strange attractor," which means, Pickover says, "It has some structure even though it's very irregular." All strange attractors are fractal.

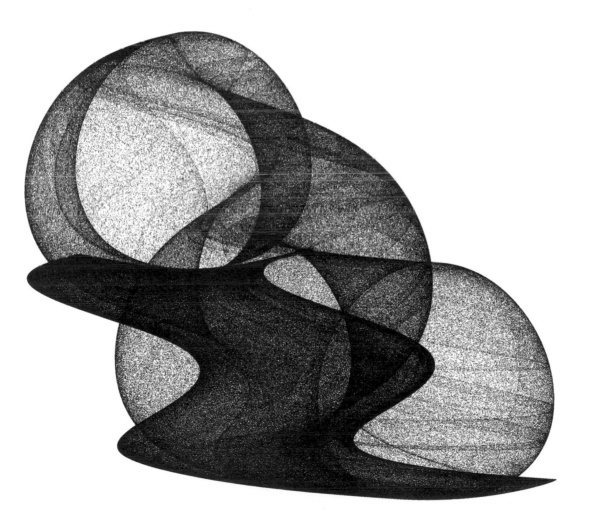

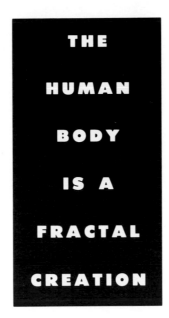

**THE
HUMAN
BODY
IS A
FRACTAL
CREATION**

•

If you like fractals, it is

because you are made of

them. If you can't stand

fractals, it's because you can't

stand yourself. It happens.

—HOMER SMITH,

computer engineer, of

Art Matrix.

•

The traditional medical model of the human body portrays it as an organic machine. Clocklike rhythms such as the beat of the heart tick away until the machinery wears out. Like an understructure of ball joints and hinges, parts of the skeleton can be repaired, even replaced. The mechanical model shows the nervous system as a telephone exchange or, in more recent high-tech metaphors, a "wetware" computer circuit board.

This image of the body sharply contrasts with the one being crafted by a new generation of scientists who have begun portraying our physiology as a holistic activity laced with fractals and chaos. Fractal geometry describes structures full of spaces and surfaces that wrinkle, branch, and fold into self-similar detail at many scales. These kinds of structures and surfaces are found everywhere in our bodies. Consider the classical picture drawn by Andreas Vesalius, the father of modern anatomy:

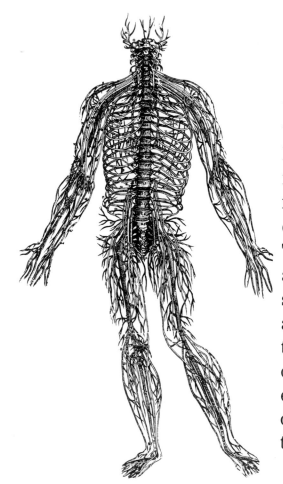

Medical textbooks are chock full of pictures like this one—drawings depicting everything from the cascade of ever smaller blood vessels feeding the heart, to the densely packed, multiple-scale branching of the entire circulatory system. The lymph system, the small intestine, the lungs, muscle tissue, connective tissue, the folding patterns on the surface of the brain, the calyx filters in the kidney, and the design of the bile ducts—all show irregular self-similar scaling. This fractal design vastly increases the surface area available for the distribution, collection, absorption, and excretion of a host of vital fluids and dangerous toxins that regularly course through the body. The intricate fractal pattern of neurons constitutes an incredibly sensitive and efficient network for information processing. Because each of the body's fractal-shaped structures is redundant and irregular, parts of fractal systems can be injured or lost with relatively minor consequences. Fractals make the body flexible and robust.

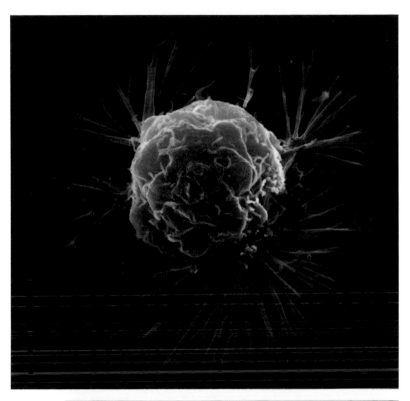

A malignant fractal. This breast cancer cell is a wrinkled, branching space-filling shape that can multiply in the human body by feeding off the body's healthy fractal structures. The shapes of other pathogens are also fractal; for example, the electrical charges in the "coat" of a polio virus show a fractal design.

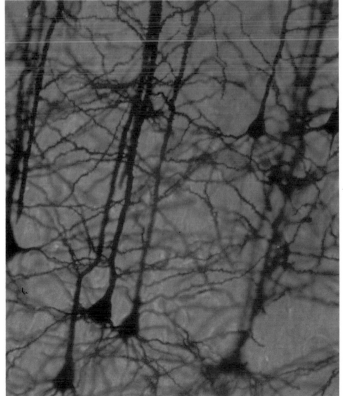

The dendritic (fractal) structure of cells in the brain is beautifully caught in this photograph by a team from the Fidia Research Laboratory in Padua, Italy. The fractal geometry of the brain allows the three-pound ball of cells in our skulls to pack an enormous, variable surface area into a small space. As a separate structure, each brain cell is free to render a unique response to stimuli, yet through the cell's branches it is compelled to participate in a network that unites it through feedback with the entire brain. In the spaces between brain cells, other fractal networks deliver the oxygen, nourishment, and hormones necessary to keep neurons firing. In all, the fractal geometry of the brain gives it a flexibility and complexity no microchip technology has begun to approach.

Inside these layers upon layers of our bodies' fractal structures, chaotic processes reign. The conventional picture of the regular, periodic beat of the heart only holds true if physicians ignore the subtleties of their patients' electrocardiograms. In fact, when the ECG of a normal heart rhythm is plotted over time, it shows considerable small variability in the intervals between beats. When a special kind of plot, known as a phase-space plot, is made of these intervals, instead of mapping out the neatly circular patterns characteristic of a regular, periodic rhythm, the pattern that emerges shows the spiderlike shape characteristic of a strange attractor. Here are phase-space plots of two normal heartbeats:

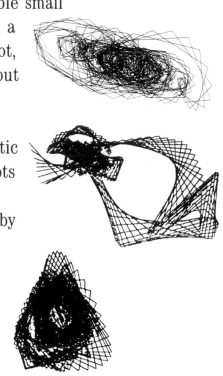

Strange attractors are fractal patterns made by a dynamical system exhibiting chaos. Recent research has suggested that chaos is normal in the heartbeat, and contrary to expectations, it is the sickest hearts whose beat often looks most periodic when plotted. Here's one of a patient who died of a heart attack only eight days after these cardiac measurements were taken.

The rhythm of this heartbeat has lost variability, its background chaos, according to Harvard Medical School cardiologist Ary Goldberger. Goldberger thinks that spotting a reduction of the heart's background chaos will help diagnose heart pathologies.

Researchers are becoming increasingly aware that pathology is related to loss of "natural" background chaos in the body. While an epileptic seizure may look from the outside like an attack of chaos, from inside the brain it is an attack of abnormally periodic order, according to Agnes Babyloyantz at the Free University of Brussels. The seizure destroys the brain's natural chaos and replaces it with the spasm of a "limit cycle." Similarly, levels of white blood cells fluctuate chaotically in healthy people but oscillate cyclically in certain patients with leukemia. Research also indicates the immune system's method for making

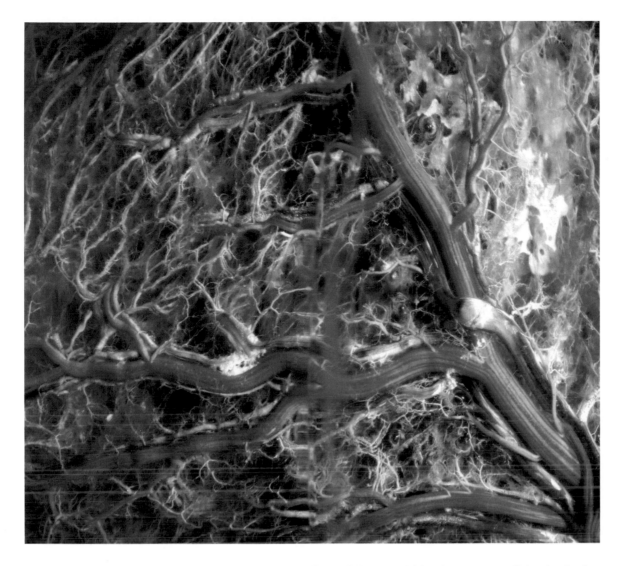

The fractal geometry of the circulatory system allows delivery of blood to every cell in the body. Typical of fractals, the branching shape of the blood vessels appears the same when you examine its detail at smaller scales.

antibodies may involve chaotic activity, and Goldberger thinks that the tremors of Parkinson's disease may arise from a loss of normal chaos in the neurological system. Even aging itself may be the result of a "loss of variability," he says, "a decrease [in] the dimensionality or degree of chaos."

Chaos in the body is caused, in part, by the constant feedback occurring as the different "parts" of these highly complex systems interact with each other. The feedback involves time delays. These build up as the feedback goes on so that the movement of any system is always undergoing a subtle shift. The variability that results from this nonlinear feedback also gives the organic system a "plasticity essential for coping with the exigencies of an unpredictable and

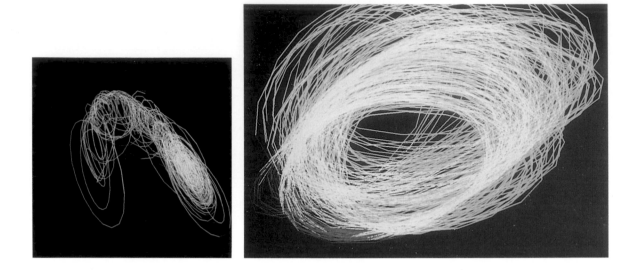

What goes on in populations of brain cells is a highly sensitive, controlled chaos, according to researcher Walter Freeman and his colleague, Christine Skarda. Shown here are the results of a computer model of the activity inside the human olfactory bulb, the processing site in the brain of our sense of smell. The model is based on actual experiments done with rabbits.

According to the two scientists, the cells in the brain's olfactory bulb are continuously firing in a chaotic pattern. A plot of their group activity is shown in the frame at the left. The curious shape of this plot reveals that though the firing pattern of each individual cell is completely unpredictable, the cells' group activity falls within a certain range. This overall chaotic shape of the activity is called a "strange attractor." Paradoxically, the chaos taking place within the strange attractor enfolds a hidden order that becomes apparent the moment we get a whiff of something.

When receptors in the nose are stimulated by a scent, they relay their excitation to the olfactory bulb. The moment that information is introduced into the bulb, it is amplified by the underlying chaos, which suddenly changes shape and self-organizes, as is shown in the frame on the right. This wiry coil picture of bulb activity is also a strange attractor, but one with more order. Freeman and his colleagues have found that the olfactory bulb exhibits a distinct self-organized strange attractor for each scent. These attractors suggest that our "memory" for the scent of roses is an implicit order embedded in the moment-to-moment chaos of our brain's electrical activity—just waiting to be called forth by a stroll past a florist. By making use of chaos, the body's systems achieve a diversity of response that no mechanical, cyclical order could give them.

Strange attractors like these are a fractal record of chaotic activity and the powerful order in chaos. Scientists are beginning to discover that the brain is interwoven with strange attractors. To see another pair, turn to the *Introduction.*

changing environment," says Goldberger. In a *Scientific American* article, he and two colleagues argue that while "the conventional wisdom in medicine holds that disease and aging arise from stress on an otherwise orderly and machine-like system—that the stress decreases order by provoking erratic responses or by upsetting the body's normal periodic rhythms"—it turns out that in fact "irregularity and unpredictability . . . are important features of health."

Our fractal body also defines us as individuals: Notice how the two "normal" heartbeat strange attractors you just looked at are quite distinct. Now consider the idiosyncratic speech patterns made by each of three speakers saying the sound of "oo" in "boot." Clifford Pickover of IBM mapped each of these patterns on a kind of kaleidoscopic mirror in order to make it easier to detect the similarities and differences between them.

Pickover envisions a time in the near future when we can enjoy pictures of our internal chaos he calls "biometric art." Here is an example of that art from parameters characterizing blood vessel patterns in Pickover's right and left eyes. He plugged the parameters into a mathematical equation used to model chaos, and it generated on his computer screen patterns that suggest the irregularity and regularity that is our nature.

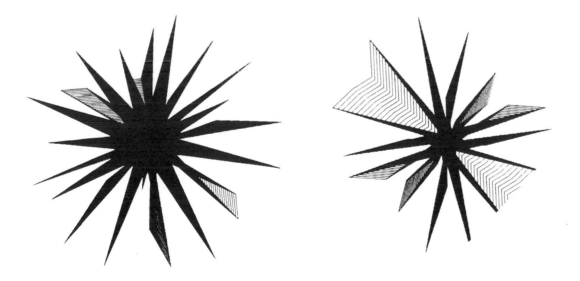

Each of our bodies is a unique signature of chaos. Even in the folds and wrinkles of our faces when we enter the world and when we leave it, we are fractals.

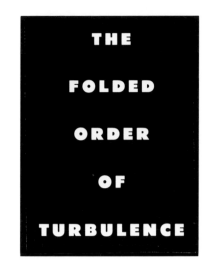

THE FOLDED ORDER OF TURBULENCE

•

What from the magnitude of the shock might have been a column of water running upright in the dark butted against the ship, broke short and fell on her bridge crushingly, from on high, with a dead burying weight. A flying fragment of that collapse, a mere splash, enveloped them in one swirl from their feet over their heads, filling violently their ears, mouths, and nostrils with salt water. It knocked out their legs, wrenched in haste at their arms, seethed away swiftly under their chins; and opening their eyes they saw the piled-up masses of foam dashing to and fro amongst what looked like the fragments of a ship.

—Joseph Conrad, *Typhoon.*

•

Shimmers and coils of heat rising from hot pavement, thunderheads boiling over the horizon, an oil slick, the spreading cloud of cigarette smoke at the next table, soup bubbling in a pot on the back of the stove: Turbulence is everywhere around us.

Painters and poets have long admired its subtlety and power: a brook crashing down a mountainside, leaves swept around in the wind, clouds in a sunset. In everyday life we count on turbulence to bring rain to the garden and the sound from a flute. But we also fear its wrath in an airplane or on the open sea. Turbulence is the quintessential symbol of chaos. When ancient Chinese painters depicted creation, they showed dragons emerging out of a turbulent whirlwind.

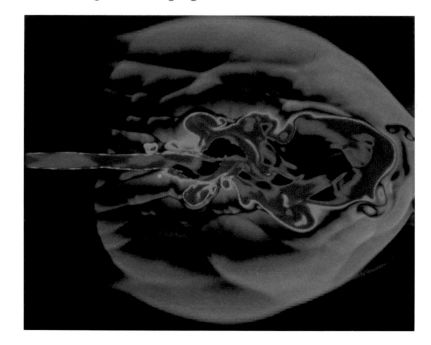

Astrophysicist Michael Norman who computed this image of turbulence on the supercomputer at the University of Illinois, Urbana-Champaign, calls this a "galactic garden hose." Though the nonlinear equations used for such computations are relatively simple, when they are calculated, their behavior becomes so ferociously complex that it takes considerable computing power to handle them. The result in this case is a model of the turbulence found in intense streams of particles emitted from the interior of galaxies. Some of the galactic jet streams are a million light years long. The jet depicted here streams from the galaxy and then flails chaotically as it hits denser matter, as if a garden hose had been sprayed into a tank of Jell-O. Modeling gives scientists the opportunity to see how well they understand the jets by comparing the behavior of their models with astronomic observations.

Oil leaking from the grounded tanker *Argo Merchant* in December 1976 unravels into the ocean. The ship was spilling about 40,000 gallons per hour at this point. The constant turbulent action of the ocean disperses oil spills so quickly that containing them entirely is impossible.

For over a century, however, scientific analysis of turbulence—that is, knowing the precise conditions that cause it and the process by which it develops—has been one of the most intractable problems in classical physics. Scientists want to analyze turbulence so they can predict and control it. Understanding turbulence would help them to design bridge pilings to resist wave action, pipelines that flow smoothly, and artificial hearts that don't inadvertently swirl blood into clots. In 1932 an eminent British scientist told a meeting of the British Association for the Advancement of Science: "I am an old man now, and when I die and go to Heaven there are two matters on which I hope for enlightenment. One is quantum electrodynamics, and the other is the turbulent motion of fluids. About the former I am really rather optimistic."

Major problems of quantum electrodynamics have indeed been solved, but the classical equations developed over a century ago to describe the growth of turbulence in a gas or fluid defy even today's most powerful computers. These equations have terms for the ratio of a fluid's mass and velocity to its thickness, but because the formulas also contain nonlinear terms, the equations keep feeding into each other. Values twist and tangle and small errors in the calculations multiply so rapidly that the results become useless. The chaologists have gained some headway in the study of turbulence, however, by plunging headlong into the chasms of its unpredictability.

The unpredictability of turbulence exists because dynamical systems made of liquid or gas are hypersensitive. A fluid system is easily folded back on itself, and its folds can grow quickly—with patterns that are as unexpected as the folds of a piece of paper suddenly balled in your hand. A few grains of ice on the wing of a jet liner, for example, can cause a wrinkle in the air current that feeds back upon itself, multiplies, and spirals to create turbulence that may be great enough to cause a plane crash.

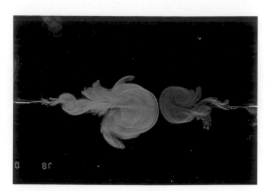

Another reason turbulence is so hard to analyze is that it takes place on many scales. Magnify a small-scale portion of a picture of a babbling brook and it looks similar to the larger-scale image; there are folds within folds within folds. At the same time, turbulence, like other forms of chaos, is paradoxical: in the midst of its disorderly motion, vortexes may appear and remain stable while the disorderly current boils on around them.

Applying the ideas of chaos theory to turbulence, scientists have discovered rules governing transitional points

Turbulence arises from a holistic folding process which makes the flow of a liquid or gas increasingly complex—as demonstrated in these wind tunnel pictures made by the Office National d'Études de Recherches Aérospatiales in France. The wrinkling and crinkling of the strands of gas or fluid as it folds is fractal.

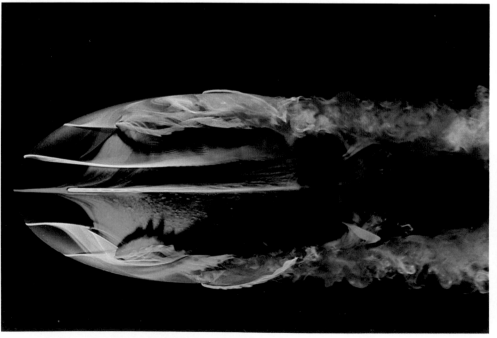

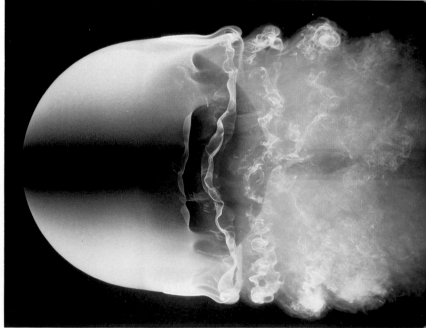

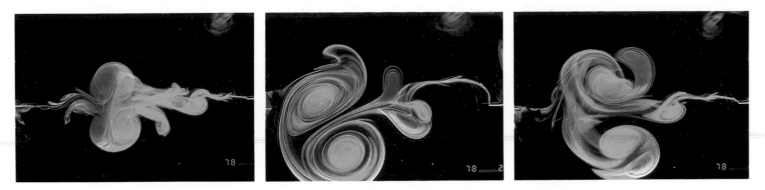

The dance of turbulence in these four frames depicts one of the paradoxes of chaos. To get a handle on turbulence, labs across the world devise methods for studying it under controlled conditions. Here scientists at the Institute of Meteorology and Oceanography at the University of Utrecht in the Netherlands squirted turbulent bursts of colored liquid into a tank of salt water and watched as the bursts collided and organized themselves into a two-headed vortex. Disorder has danced momentarily into order—but order of an essentially unpredictable kind.

from smooth flow to rough flow and are beginning to understand in principle how the feedback folding process takes place. Using nonlinear equations that are simpler than the classical equations developed in the last century, chaologists have even been able to craft realistic computer images of turbulent flows. The images will get more detailed as computers get more powerful, but because of the nature of chaos it is unlikely that the riddle of turbulence will ever be solved enough to make detailed predictions possible.

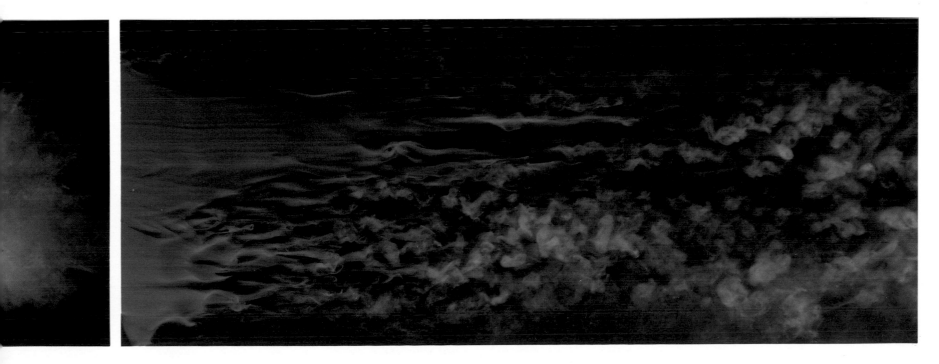

Perhaps one of the reasons turbulence has fascinated artists is that its subtleties mirror the subtle shifts in our own psychologies and moods. Here Lawrence Hudetz captures several moods of turbulence in his photograph of McCord Creek in Oregon. Leonardo da Vinci was one of many artists across the centuries inspired by the mysteries of turbulence. For da Vinci turbulence was apocalyptical; he believed the world would end in a cataclysmic deluge.

VISUALIZING CHAOS AS A STRANGE ATTRACTOR

•

I have not spoken of the

aesthetic appeal of strange

attractors. These systems of

curves, these clouds of points

suggest sometimes fireworks

of galaxies, sometimes

strange and disquieting

vegetal proliferations. A realm

lies here to be explored and

harmonies to be discovered.

　　　—DAVID RUELLE, one of

　　　the world's foremost

　　　authorities on chaos

　　　and dynamical

　　　systems.

•

Scientists have a passion for plotting things. Perhaps it started with the great French scientist Rene Descartes and his British successor, Isaac Newton. Descartes and Newton laid a grid over the universe and proved that everything that moves can be measured and specified by sets of coordinates. Scientists have been carefully plotting things ever since.

When scientists took to plotting the movement of classical dynamical systems—systems chugging around at a measured pace—they often filled their plotting space with a neat, smooth-looking shape called a torus. A torus in its three-dimensional form (toruses can exist in many dimensions) looks like a well-formed bagel or an American doughnut with a hole in it. Scientists plotting well-behaved dynamical systems such as planets in their orbits, or oscillating electrical devices, learned they could wind imaginary wires around the surface of an imaginary torus to indicate the orderly functioning of these well-behaved systems. For example, the orderly movement of a planet in orbit can be depicted as a line that winds around the surface of a torus, repeating the same path but shifted slightly with each circuit. Then the chaologists came along.

The chaologists wanted to study dynamical systems in a wilder state: the chaologists wanted to measure systems as they broke down, disintegrated, came apart, or fluctuated unpredictably and transformed themselves. When the chaologists plotted toruses for these systems, the results were strange—doughnuts riddled with twists, folds, and curious internal shapes. Let's look at an example:

One of the chaotic dynamical systems studied involved some peculiarly empty orbits in the belt of asteroids between Jupiter and Mars. A Soviet scientist, Andrei Kolmogorov, proved a theorem that showed there was chaos occurring in these orbits as a result of the friction and resonance set up by the combined effects of the motions of Jupiter, the Sun, and Mars. Imagine the choppy wash that occurs when large ships and powerboats pass each other and their waves collide. If you were sitting in a rowboat caught in one of those places where the waves met, you'd feel the chaos. The mathematics of the Kolmogorov theorem can be plotted out as a torus; cut it open, and it provides a visual picture of the choppy mess that occurs when you take into account several contending planetary motions.

Technically this torus is called the Vague Attractor of Kolmogorov or VAK, for short. The acronym is apt, since Vak is the name of the goddess of vibration in India's ancient holy text, the *Rig-Veda.* The VAK torus shows that the chaotic asteroid orbits exhibit some regular motion—indicated, for instance, by the red arrows winding around—and otherwise a lot of wobbling and weaving. Any chunk of matter that had the misfortune to wander into such an enchanted, hellish zone would oscillate drunkenly until it was eventually expelled. The Vague Attractor of Kolmogorov is the first in our gallery of strange attractors. The phrase "strange attractor" was coined as a kind of attempt at scientific humor. Classical, "regular" systems such as the orbits of Mars or Jupiter can be plotted as smooth and regular-looking shapes like the torus. Scientists say the movements of these classical systems are "attracted" to those orderly-looking shapes, which are abstract portraits of their orderly behavior. But the movements of chaotic systems seem attracted to strange shapes like the messiness inside the Kolmogorov attractor. One of

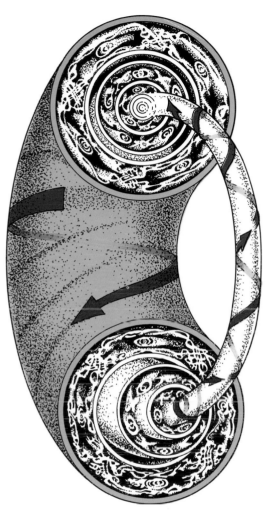

the really strange things about strange attractors is that they do have a predictable overall form, but it's a form made of unpredictable details.

By using equations to follow one or more of the variables of a chaotic system as it changes and moves, scientists can plot out a strange attractor that portrays the system's activity. To create these pictures of strange attractors, the chaologists' equations are calculated to an output and then the output becomes an input as the equation is calculated again. This mimics the kind of accelerating, amplifying feedback that goes on in real chaotic systems—the factor that makes these systems constantly transform themselves. Think of the weather or a mountain stream. The system's holism (the fact that every movement in the system in some way affects every other movement) is responsible for its chaos (unpredictability). At the same time the weather is constantly changing, it also stays within the boundaries of what we call the climate, just as a turbulent stream stays within the boundaries of its banks. But even strange attractors can convulse and change their basic shape if the system is perturbed enough, just as a heavy rainstorm can make a stream burst its banks and take a new course. Climatologists worry these days that the weather's strange attractor (the climate) may one day change its shape as a result of the industrial perturbations caused by human beings.

But barring such earth-shattering metamorphoses, with each iteration of feedback the chaotic system folds into (or explores) a new region of the space inside the tangled outline of its strange attractor. In fact, the boundary of the attractor itself is constantly being redrawn and complexified, as the iterations loop their way into new dimensions. Here's another picture of a strange attractor.

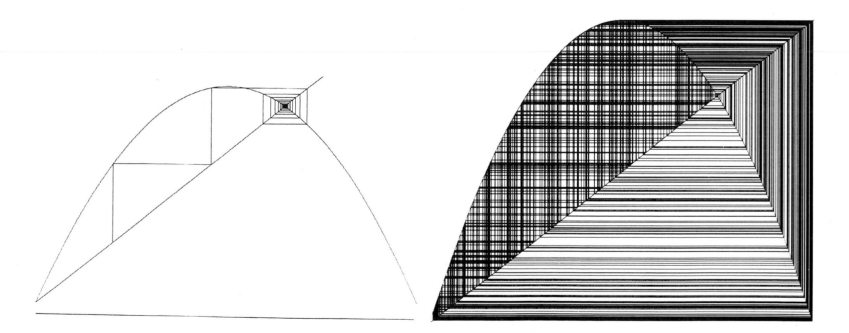

It's called a "cobweb diagram." You don't have to know how to read it to see this as a picture of chaos.

Volcanic eruptions like this one in the Philippines are accompanied by tremors which, when plotted, reveal an underlying strange attractor. This particular one is called a Rösseler strange attractor. Curiously, the Rösseler has been found to apply to quite a different kind of dynamical system from volcanic tremors. It also shows up when scientists plot the Belousov-Zhabotinskii chemical reaction. Here the chaotic bonding of the chemical reagents self-organizes to create highly structured spiral-like forms. (See **Self-Organization**.) In other words, the Rösseler attractor plots the transition from order to chaos, but it also plots the transition from chaos to order.

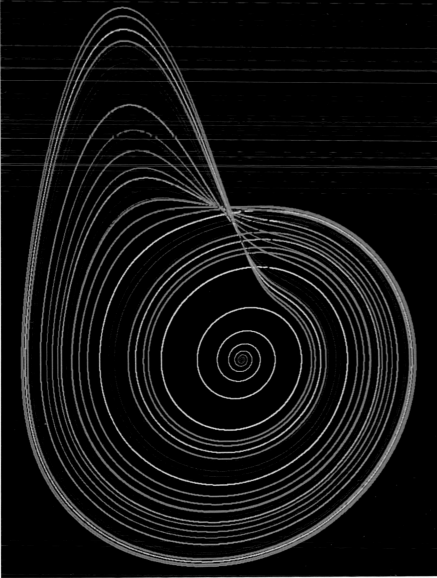

A view of the Rösseler strange attractor.

This fanciful torus, computed by Cliff Pickover at IBM, might be used to describe a smooth-running, "classical" dynamical system, or it might be a holiday wreath.

This is a slice through a section of a chaotic torus called a Ueda strange attractor. The section shows the torus continuously folding in on itself, like a pastry chef stirring colored dyes into cake batter. The Ueda attractor shows up when scientists plot equations that model dynamical systems such as the oscillation of an electromagnetic field within a ring-shaped cavity or the rise and fall of certain types of predator and prey populations. The gold-colored region indicates a type of behavior more frequently exhibited by the system. The crimson region denotes less frequently exhibited behavior. Magnifying a small-scale portion inside a strange attractor reveals shapes similar to those seen on the larger scale. Because of the self-similar way they fill space, strange attractors are fractal. They are fingerprints of the chaotic dynamical systems they plot. It is said that the resemblance of Ueda attractor to the ancient Chinese yin/yang symbol for change is purely accidental.

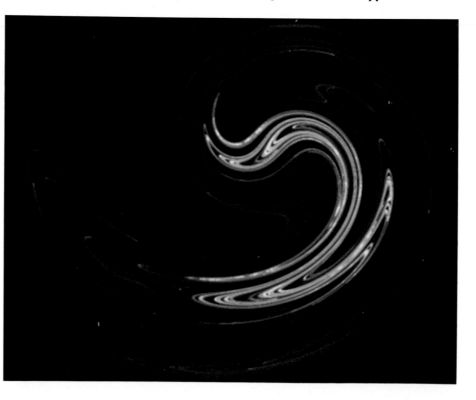

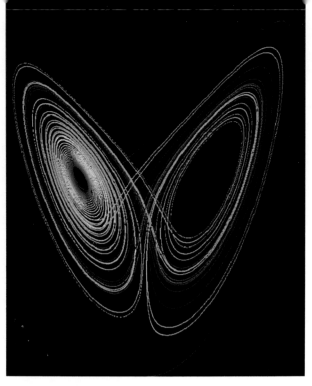

This is one face of chaos, the butterfly mask of unpredictability discovered in the early 1960s by one of the first chaologists, Edward Lorenz. When Lorenz plotted several variables describing the movement of a weather system, he discovered that very small differences in the initial daily weather data that he input into his model would make a very large difference in a long-range forecast. That meant that a meteorologist calculating two weather projections where he started with similar but not identical data would end up with two entirely different long-range forecasts. By iterating the equations of his model to make a plot, Lorenz produced this strange attractor, which is a fractal portrait of the essential unpredictability he had found in the weather.

The repeated folding of the system onto itself (in other words, the continuous interaction of variables such as temperatures and pressures) is represented by the fold between the two "eyes" of the attractor. The recurring shapes that circle around the eyeholes indicate that the weather is unpredictable but self-similar: high- and low-pressure gradients, temperature variations, and other factors exist on every scale, from global weather patterns to local variations between the front and back yards of your house.

Strange attractors like this one depict a system whose behavior never repeats itself and is always unpredictable and yet, paradoxically, always resembles itself and is infinitely recognizable.

The chaologists have discovered all kinds of wonderful strange attractors: portraits of the order in chaos. IBM scientists found a new one when they plotted the activity of two barium ions caught in an electronic "trap." By varying the strength of the energy they were using to trap the ions, the researchers could watch this relatively "simple" system display a veritable carousel of behavior. At one frequency, the system would freeze or "crystallize" the ions so that they would hover next to each other catatonically. Higher frequencies set them bouncing randomly off the energy walls of the trap. In the midst of this chaos a slight change in frequency could set them oscillating or dancing in an

This gossamer piece of abstract art is called a period doubling plot or logistic map. It's another kind of strange attractor. The dynamical system modeled here might be commodity prices or populations of gypsy moths. Reading from left to right, we can watch the system going to pieces.

As some variable is increased, for example, the number of trees gypsy moth caterpillars like to eat, the population jumps and more eggs are left behind to hatch the next year. That year, however, there are too many moths, and they outrun the food supply, so the following year there is a die-off. The population oscillates between two values—one year it's high, the next year it falls off. Increase the starting food supply still further and the population goes into a four-year cycle; make another increase and the cycle goes to eight years; and so on. Set the food supply high enough, and the population rate cascades into chaos, making it impossible to determine what the numbers will be in any year. Scientists have discovered laws (in the form of ratios) governing when the transition from two cycles to four to eight and up to chaos will occur in these sensitive systems. The period doubling plots shows that as more energy is injected into chaotic dynamical systems they fold into themselves and wrinkle up, becoming increasingly intricate. The wrinkling and folding reveals their fractal nature.

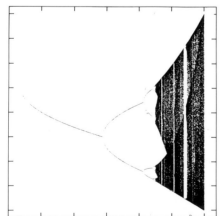

Notice the two black lines in the plot. These are windows or clear spaces smack in the midst of advancing chaos where for a short time—a few years—the population of moths might rise and fall in an apparently cyclical way. But then the pattern dissolves into chaos again. Notice the miniature plots inside these windows:

These are indicators of the chaotic system's self-similarity. A small-scale period doubling toward chaos broods in the middle of the narrow window of order. In terms of the gypsy moth example, the effects of this small-scale self-similarity would be too subtle to be detected.

The period doubling plot shown here in color represents an incomplete portion of the period doubling cascade and was selected for its aesthetic appeal. Klaus Ottmann, an art curator who in 1989 organized the first exhibition of paintings, sculpture, and computer graphics on chaos in North America, said it was a color period doubling plot that made him realize the deep artistic possibilities of fractals and chaos.

orderly pattern, "phase locking" as scientists put it. The strange attractor the researchers plotted here is a cross section through a chaotic torus.

The hexagonal region near the black center plots the area near the frequency where the ions self-organize into a phase-locking order. The spiral arms plot frequencies where the ion activity is spiraling off into chaos.

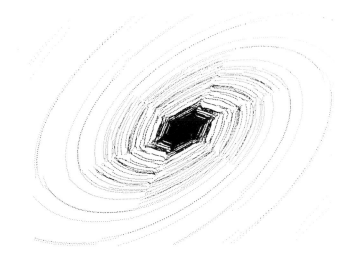

More and more, scientists are spending time in a strange space that Newton and Descartes would have never imagined.

THE ART OF ABSTRACT IMAGES FROM FRACTAL MATH

•

Dragons fight in the meadow.

Their blood is black and yellow.

Image from the I

Ching, the Chinese

Book of Changes,

depicting dragons,

which represent order,

faced in a conflict that

leads to chaos.

•

In 1918, the great mathematician, philosopher, and logician Bertrand Russell remarked that "mathematics, rightly viewed, possesses not only truth, but supreme beauty—a beauty cold and austere, like that of sculpture." One wonders what Russell would say about the sculpturelike shapes of computer-generated fractals. The art made from fractal equations is beautiful but hardly "cold and austere." Its patterns crackle with color and detail, like an excited nervous system or a fireworks display. In Russell's day one had to be a mathematician to "see" math's sculpture. Nowadays, thanks to fractal geometry, even people with severe math phobia can experience in tangible form what Russell meant when he insisted that mathematics possesses "supreme beauty."

The eerie, carnival-like creatures of mathematical fractals have sprung up on the covers of science books around the world. These images are in fact computer-made portraits of sets of equations that have in common the characteristic of *iteration,* a mathematical form of feedback. When a value is put into one side of a fractal equation and the equation is computed to a result, the result is then inserted back into the equation and the equation is computed again. The new result is then reinserted and the equation is iterated (rerun) once more. Some starting values, when they are plugged into an iterative equation cycle, explode toward infinity, others fluctuate, some don't change very much. Equations that undergo sudden unpredictable behavior when iterated are *nonlinear.*

Nonlinear equations are extremely sensitive in some regions of values, and those values mark the borderland between mathematical order and chaos. When these borderland values are plotted on the computer screen using color, presto, the equation's dynamical activity reveals a region that is brilliantly fractal. Some fractal equations have been invented to model real chaotic systems, others to probe the chaos that lies hidden in mathematics itself.

As they plot their fractal equations, many scientists are finding themselves drawn toward the ancient aesthetics of art. One reason may be this: Complex dynamical systems—that is, systems undergoing constant change because they have many "parts" feeding back into each other—are holistic in the sense that everything in these systems affects everything else. Both dynamical systems and mathematical fractals exhibit self-similarity in that their very different-sized "parts" subtly reflect each other. Self-similarity and an implied holism are two vital perceptions in the age-old aesthetic artists have employed to make forms that mirror, mimic, or metaphorically invoke the cosmic mystery.

Science's drift toward art peppers the reflections of three researchers who, in different ways, create portraits of fractal equations on their computers.

Scott Burns, an associate professor of engineering design at the University of Illinois, Urbana-Champaign, studies a curious piece of fractal mathematics called Newton's method. The method—named after its inventor, Isaac Newton—is a shortcut for finding the roots of a *polynomial* equation (an equation with several terms). Starting with a guess at a root's value, the mathematician plugs the guess into the method's formula and iterates, watching as each iterative loop of the method changes the guess so that it gets closer and closer, "converging" toward some fixed number which is one of the polynomial's roots.

However, if the starting guess happens to be a value that lies on the boundary regions *between* roots, then Newton's method turns into chaos. By plotting the different starting "guesses" and coloring them according to whether iterating them makes the result converge toward one of the roots, fly off into infinity, or lie in the boundary area, Burns obtains a fractal picture.

Newton's method is used by engineering designers to solve problems such as finding the best size timber for a structure composed of two columns holding up a supporting beam. An optimal engineering design is one that makes the most efficient use of materials and is safe. To locate the possible optimal designs using Newton's method, the engineer shapes an equation that expresses such factors as the levels of stress, strengths of timber, and how much different-size timbers bend. The engineer then takes a guess ("columns 2 inches square, beam 4 inches square," for example), and iterates the guess using Newton's method. In most instances the engineer finds a solution. In some cases he finds the fractal shape of a mathematical chaos.

Burns, who works on Macintosh personal computers, shows some of the fractals he creates using Newton's method at craft fairs and galleries. "I do that so I can get in there and talk to people about math and art. I find that people are fascinated by this stuff."

Burns says his mission is to convey the beauty of mathematics because it's

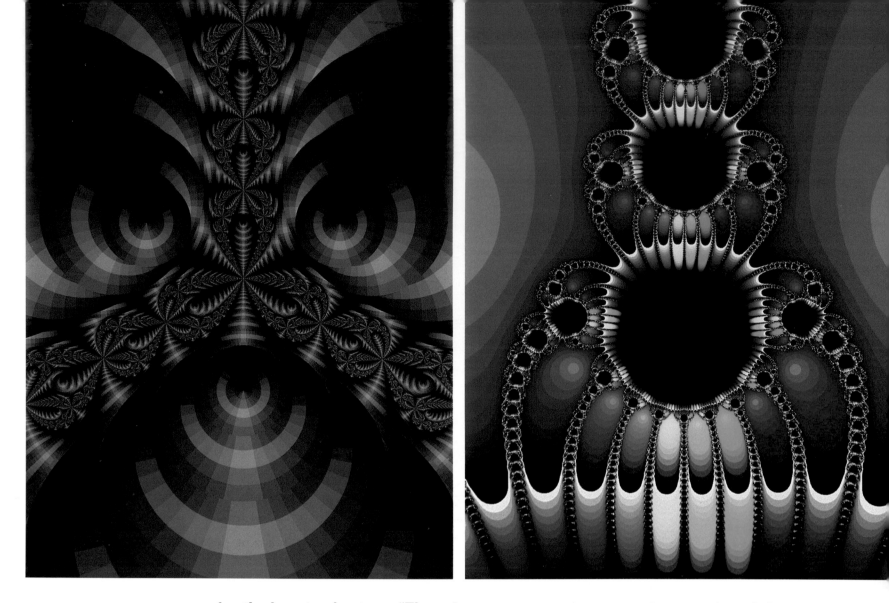

also the beauty of nature: "These images represent a personal expression of the hidden beauty. You may question, is it art? In some ways these images may be thought of as paint by numbers on a grand scale. I don't presume to take artistic credit for the many shapes and patterns. They occur naturally in the mathematics." But he does make choices: the color palette, when to shut off the iterations. "I can focus the picture, but I don't really have control over what it is that's being seen."

The first image here is an example of Newton's method applied to an equation that has three roots. The equation's roots are the tips of the "umbrellas" in the picture. "The chaotic region is where the forms get smaller and smaller and seem to come together. My wife calls this one 'a pregnant woman.'"

The second image is a close-up of a boundary region of an equation with ten solutions. The pinkish areas to the left and right are locations of two of the ten solutions. The black hole represents a region of points (starting guesses) that hadn't begun converging toward any of the ten solutions by the time Burns stopped iterating them. "Everywhere you see a black hole, it obscures a region of chaos. The boundary everywhere is covered by black holes."

Mario Markus is a physicist at the Planck Institute in Dortmund, Germany. Using his computer screen as electronic graph paper, he plots the twisted chewing-gum complexity of a series of equations that describe the transition zones from order to chaos. These equations can be used to model real systems which have complicated interactions (so-called dynamical systems) such as the flow of energy in electrical circuits, and turbulence in fast-flowing water.

Markus's plots are looming, surreal, perhaps troubling evocations of the infinities that lie everywhere concealed inside nature's movement. The deep blue background in each plot describes the dark domain of total chaos. The infinitely intricate shape in the foreground is a forbidding fractal creature that breeds and lives in the region of transition. These are borderland creatures of order. The shadowy organs and veins inside the creatures represent "superstable" areas which resist change. Notice the complex self-similar copies of the large form that appear on smaller and smaller scales—the characteristic sign of a fractal. In the blue sea of chaos these shapes represent small islands of order that lie between larger mainlands of order. Markus says, "This implies a never-ending appearance of such regions, separated by chaos, upon successive picture magnifications. It is thus not always possible to say: 'The system is chaotic in the parameter-interval so and so,' because any interval of chaos may contain intervals of order at higher levels of resolution. This means for me that the question, 'Is God playing dice or not?' [in other words, is the universe ruled by chance or by predictability] cannot be answered without performing the impossible task of thoroughly exploring the filigreed maze of these fractals."

Markus confides that making his plots has "brought me a new form of art and has made me feel like an artist. Surely, one could make the objection: these pictures were produced by the computer and I just had to press a few buttons. However, this objection could also be made about photography. It can be said that one only needs to look through the camera and press a button. The reason photography is considered an art is that a good photographer does a lot more than push a button: He chooses an object, an angle, a lens opening and time to shoot out of millions of possibilities. Furthermore, he can manipulate darkness and contrast in his lab. A photographer thus has many degrees of freedom with which to express an emotional state within a high-dimensional space of control parameters.

"Much like a photographer, I have found myself moving in such a huge space when producing my fractal images. The parameters I control are degree of zooming, window, horizontal and vertical scales, colors, and sometimes a third dimension according to some intensity level. An even greater diversity is possible when one starts to change and choose the coefficients of a formula. Truly one can say that equations can be considered here as new types of painting brushes."

One interesting wrinkle to contemplate about Markus's pictures. Turn the page ninety degrees to the right and examine the image. Though what you have just done would make a scientist cringe, isn't the image somehow more "aesthetically significant"?

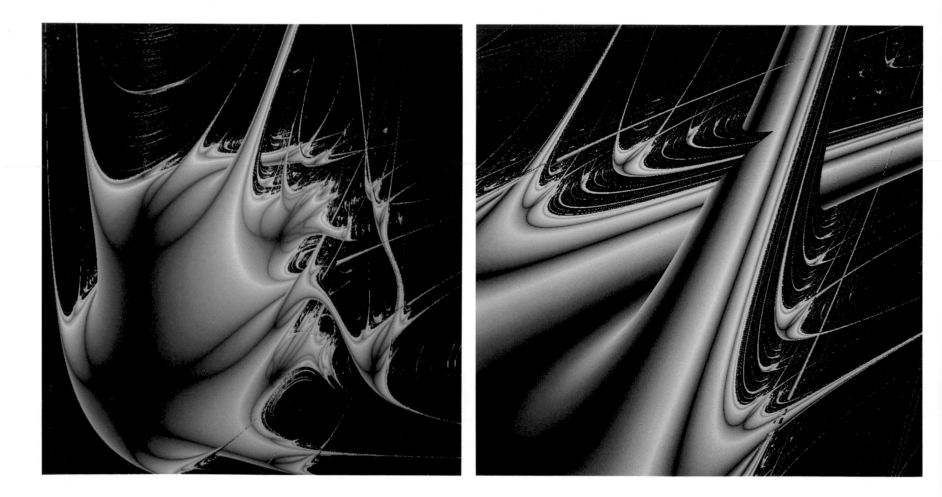

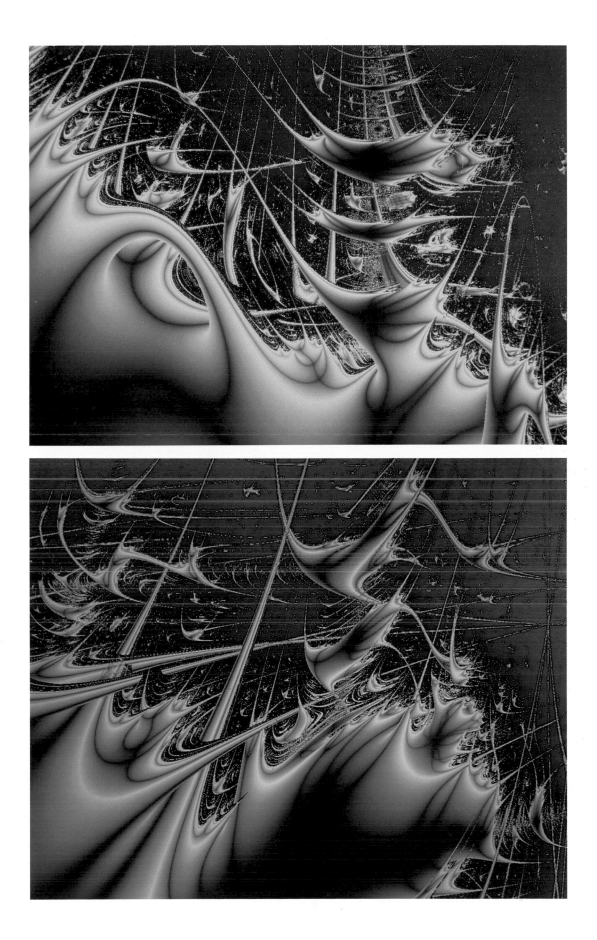

Cliff Pickover is a member of the Visualization Systems Group at IBM's Thomas J. Watson Research Center in Yorktown Heights, New York, and author of *Computers, Pattern, Chaos and Beauty* and *Computers and the Imagination.* The small room where he often works is jammed with computer screens and consoles. He types in some numbers and jumps to another screen which sparks into life with a gray, filigreed structure called a Julia set, a mathematical object related to the famous Mandelbrot set which adorns so many book covers.

The Julia set is actually a mathematical construct located in a thicket of numbers called the complex number plane. To "find" the fractal outlines of the Julia set, Pickover's computer screen becomes a piece of ultrafine electronic graph paper. All the pixels, or points, on the screen—and there are over a million pixels at high resolution—are like points marked at the intersections of the lines on a graph paper. Pickover's powerful computer "tests" each point (number) in an area of the complex plane by applying an iterative equation to it and recording how fast the value expands. If the value remains stable, he assigns it one of 255 possible colors (usually black); if the value soars quickly to infinity he assigns another color; if it lifts upward at slower rates, he assigns a color for each rate.

This morning in order to demonstrate his technique, Pickover begins with the Julia set in a gray scale (the first frame in the sequence above). Each shade in the picture represents a collection of points expanding at approximately the same rate. The black areas mark the best-behaved numbers, the most stable points—points that lie within the Julia set itself. Pickover also performs some plotting "tricks" to get special effects like the hairlines arcing down into the set's boundary.

His first color application is what he calls a "default" palette. He pecks on the computer keyboard, and the gray scale Julia is washed with colors advancing across the screen in waves covering the beaches of the nested coastlines of numbers that lie around the edge of the set. "This one is a palette I like working

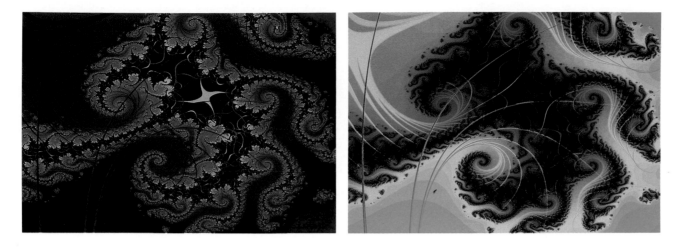

from because you know that green is next to blue, and red is far from blue. So it has some meaning to your eye." He stops on one color pattern he likes and decides to print it on a slide (second frame in the sequence here). Then he comments, "It doesn't highlight the fractal edge. In this image, the bright hot colors are the fast ones, and the greens and blues are the slowest."

To make the third frame, Pickover pecks keys and switches to his favorite color palette. "We'll highlight the structures to make something aesthetically pleasing. The aesthetic appeal of the fractal is probably created by the sharp contrast your eye sees; contrast is also useful scientifically. I'm going to start shifting the color table. There, that could be good. What do you think?"

To paint the last frame Pickover applies a "random" color table to the set. "We'll try some violets or some strange color. Now we're experiencing colors that we didn't have before. The previous one had no magenta." He pecks rhythmically at a key on the computer, and colors splash across the nested fractal shores. "That's nice," he says enthusiastically. "I like that. See, we just brought randomness into our life. You can probably make some philosophical statement about that—about the role of randomness in art."

Pickover began his scientific career in molecular biophysics and biochemistry, but he now spends all his time working with computer graphics in an effort to employ "the aesthetic side of the computer to represent biological structures and other complicated data." He says, "Many things in nature are fractals for a variety of reasons, whether it's to increase surface area or because it's easy genetically to have a fractal rule to repeat repeat repeat. These equations, even though they're purely mathematical, have that same approach, repetition leading to similar features on different-size scales."

He agrees that the patterns human beings generally find aesthetically pleasing are a dynamic balance. "Too much order and it will be like the test bars on a TV screen. Too much noise is like static. You want something in between."

As he works on the fractal forms he generates, Pickover asks everybody who

walks into the lab what he can do to improve the images. He is eager to make changes and try something new. "The computer is a tool that lets artists, mathematicians and scientists see unexpected and strange new worlds that they couldn't have appreciated before. It also lets nonartists participate in what we might call art. Art critics might not call it art, but the works I do are, to me, in the realm of art."

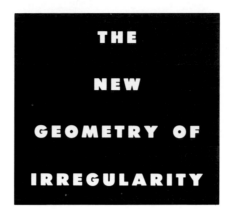

THE NEW GEOMETRY OF IRREGULARITY

•

Clouds are not spheres,

mountains are not cones,

coastlines are not circles, and

bark is not smooth, nor does

lightning travel in a straight

line.

—BENOIT MANDELBROT,

inventor of fractal

geometry.

•

Euclidian geometry idealizes forms. Triangles and squares are made with straight lines; the shapes of circles and curves are smooth and regular. Euclid's geometry defines space in terms of discrete dimensions—the zero-dimensional point, the one-dimensional line, the two-dimensional plane, the three-dimensional solid. We build our houses and cities to Euclidian specifications, and the measurements of this geometry are certainly useful for that purpose. Applied to the shapes and motions of nature, however, Euclid provides a less satisfactory grasp of the touseled, craggy, crinkly continuum of the non-human world.

Fractal geometry is idealized, too, as all mathematics must be, but vastly less so than its predecessor. It is a geometry that focuses on dynamic movement, ragged lines, and space so crumpled that it is neither line nor plane nor solid. Artists love it.

For artists, irregularity has always been a *sine qua non.* Even the obsessively rectilinear modern Dutch painter Pieter Mondrian left drips and faint wavers in his straight lines to indicate the presence of the human creator behind the abstract mathematical shapes. Michelangelo hewed his sculpture by following the grain in the marble. Nature, of course, never makes perfectly straight lines or perfectly symmetrical curves. Even the elliptical orbits of the planets wobble. Artists know that the subtle irregularity of a line, its variable thickness, embodies its energy, its significant *life.* Indeed, it might be argued that irregularity is an important feature of art and an integral part of what makes an artwork beautiful and true.

Fractal geometry moves away from quantitative measurement, which values quantifiable features like distances and degrees of angles, and embraces the qualities of things—their texture, complexity, and holistic patterning (that is, their patterning at various scales). The aesthetic—the idea of order—in fractal geometry is therefore closer to the ancient aesthetic of the artist than the aesthetic of Euclidian geometry has been. While most artists don't apply fractal geometry in any formal way to their work, most grasp fractal principles immediately when introduced to them.

The two artists whose pictures are portrayed in this section have, in different ways, taken inspiration from this new math for measuring the world.

Photographer Lawrence Hudetz's first career was as an electrical engineer. Years after he became a professional photographer, he read about chaos theory and it transformed his art.

He recalls: "I understood the Euclidian concept of the circle, triangle and square, particularly since I was using a square format. But I had always sensed there was something beyond it. I didn't quite know what or how to deal with it." Hudetz felt that chaos and fractals gave him the answer. "What thinking in fractal shapes does for me is give me another dimension. With the old geometry, I'd go into the forest and I'd be looking to line up the trees or get a certain rhythm going while the background chaos of the branches became just that, a background. But when I take the background as the real subject, then the fact that the trees are straight or not straight becomes accidental. The new geometry is a more open way of looking. This creates a subtle shift. It allows me to accept images that I might have rejected in the past because my brain was saying, 'That's not organized right.' Still, if I'm not careful, when I take the camera out, the first thing I start to do is to go back to the old Euclidian mode because it's so comfortable. We've all gotten used to seeing in that old way."

Hudetz describes himself as an artist in search of images that match what he calls his "inner fractal," by which he means his search for a texture, an inner pattern of roughness and tangledness that constitutes his sense of being in the world. He says it is a "quality of *being* that needs to show up" in the situation he wants to photograph. "I can't really say anything about it. It's there or it's not. If I try to analyze it, then the thing falls apart."

He claims that if the photograph comes out right, it will portray the exact transition point, the intersection of order and chaos. "What I want is that when you look at the photograph, there will be no way to say which direction this thing is going. Am I photographing this thicket of alders as order arising from chaos, or is it order just going into chaos?"

The first photograph here Hudetz calls his "Euclidian woods." The compositional emphasis is on the straight line. The picture has a comfortable, classic beauty. The second photograph is more disturbing. The sun-shot entwining of limbs reaches out beyond the frame of the picture so there is no clearly defined top, bottom, or sides to the composition; foreground and background are mingled.

Landscape painter Margaret Grimes discovered fractal geometry only after her painting had already undergone a major transformation. But fractals and chaos quickly helped confirm for her the importance of her new perspective.

She says, "I went through a very traumatic period where what I found exciting visually didn't fit any of the traditional forms of how to look at landscape. For example, I have a view from my house down to the water with all the underbrush, all the vines on the trees, which are killing the trees—it is overwhelming. We've got all these anesthetized ways of looking to keep us from *really* looking at nature. Of course, *not* looking is what can really kill us as a species. So I realized Things are very generalized in the traditional landscape; lines are cleaned up. There is a parklike quality."

Grimes describes painting landscapes with the new geometry as "a focus on the complexity of natural forms, the relationships—the spatial and shape relationships, and the life relationships. The formal thing is held together with patterns that are set up across the composition. Once you have that formal structure you can become very detailed in your observations without destroying your composition. So you have a shallower space, a sense of each thing in it being of almost equal significance, and that has philosophical implications: one life-form isn't necessarily more valuable than another.

"Many of the paintings take months to finish, but I attempt to retain the freshness and immediacy of the original vision—to make even a huge, complex painting look as if it happened all at once." She reports viewers sometimes get upset looking at her paintings because "they don't look as if they have any boundaries."

She says that her concern over the accelerating human destruction of the natural environment gives an urgency to her work. "In art school we were told to paint as if you were looking at something for the first time. I try to paint as if I'm looking at nature for the last time." She believes the artist should be a "shaman," uniting the viewer with the sacredness and mystery of nature, helping viewers rediscover a deep immersion in the natural world.

Like Hudetz's photograph of alders, Grime's forsythia painting dismisses comfortable triangular vanishing points of perspective, enmeshing the viewer in the riot of life.

GREAT ART'S FRACTAL SECRETS

•

Computers can make errors, of course, and do so all the time in small, irritating ways, but the mistakes can be fixed and nearly always are. In this respect, they are fundamentally inhuman, and here is the relaxing thought: computers will not take over the world, they cannot replace us, because they are not designed, as we are, for ambiguity.

—Lewis Thomas, Late Night Thoughts on Listening to Mahler's Ninth Symphony.

•

Klaus Ottmann, a museum curator who in 1989 organized an exhibition entitled "Strange Attractors: The Spectacle of Chaos" thinks there is a fractal revolution taking place in art. Careful not to call what's happening a style or a movement, he calls it an "activity":

"We might speak of a fractalist activity as we once spoke of a surrealist or a structuralist activity," Ottmann says. "Fractalist artists are both a mirror of the psychological and social state of society, and an interface. They no longer concern themselves with the mere manufacturing of objects but with the experience of fractalization." He advises, "Watch for the presence of any one of the three attributes of fractals (*scaling, self-similarity,* and *randomness*) to determine whether the fractalist vision is at work."

Indeed, contemporary artists in the United States, Europe, and Asia are rallying to a kind of fractalist's vision. One reviewer of Ottmann's show enthused: "The very simultaneity of order and disorder in the images included in this exhibition is something new. . . . Not since Stonehenge have the natural world's mysterious workings seemed so full of portent for both art and architecture."

The contemporary artists whose "activity" is fractalist claim a heritage extending back through the history of art. Indeed, the list of artists who have employed what are now recognized as fractal images would be very long: Think of Vincent Van Gogh's dense swirls of energy around objects; the recursive geometries of Maritus Escher (who said, "Since a long time I am interested in patterns with 'motives' getting smaller and smaller till they reach the limit of infinite smallness"); the drip-paint, tangled abstractions of Jackson Pollock; the detailed baroque design of the Paris Opera House; the scales of recurring arches in Gothic cathedrals; and the mountains in ancient Chinese landscapes that have the turbulent look of frozen clouds.

But now, at the end of the twentieth century, there is conscious fractalization. Art has become "a self-referential and self-reproducing system," says Ottmann. Today's artists are excited by the recognition that fractalization, in some deep sense, *is* art. However, the rise of fractals has also democratized art and posed a serious question for contemporary artists. In a symposium that followed his art exhibit on chaos, Ottmann brought together fractal imagists from science, computer graphics, and the fine arts. There, IBM's Clifford Pickover put the

Carlos Ginzburg, a Parisian artist and member of a coterie of *"fractalistes"* in Europe, calls this piece "Chaos Fractal 1985—86." Seen on a large scale, at a distance, its surface appears randomly abstract with jagged islands and edges of colors. Viewed at small scale, the piece reveals a wealth of surprising new detail of cut-out objects and patterns whose psychological, social, and relative size scales are mixed.

Ginzburg admits playfully that "understanding fractals and chaos made me more than change my perceptual experience of the world. My 'Homo Sapiens'—'Homo Faber'—'Homo Demens'—'Homo Ludens' dimensions changed definitively into a 'Homo Fractalus' one. I'm a fractal subject-fractalman." He adds, "Fractals are the scheme, the main scheme of our culture. We are now at the 'fractal state of value' and fractals show the viral proliferation of society and individuals." Asked what

he thinks about the relationship between order and chaos in nature, Ginzburg replies, "I'm really far away from 'nature,' based inside the electronic mode of information, playing the game of simulated order and chaos."

New York artist Edward Berko says that he uses the ideas of fractals and chaos "to explore the manifestation of structure in nature" and that he became interested in the theories from the viewpoint of aesthetics. "I paint in order to explore the potential of fractal geometry, to express a reinterpreted aesthetic of nature." In an essay called "On the Nature of Fractalization," Berko describes his aesthetic:

"We find strange, unnatural connections between ideas. In this circulating rush, we postulate against originality. We posit the question: Are we in a condition of infinite repetition? Infinite self-similarity? Infinite magnification of differences which is actually the sameness we thought of as difference? We contemplate the search for order within sameness, order opposing sameness, order within random behavior. . . .

"Just as the creation of a fractal structure involves the process of iteration, so the production of artistic works involves iteration. The creative process is a system wherein the output eventually becomes part of the input. In this way, the process of making art becomes self-similar, self-referential and an iteration of itself."

Berko calls this piece "Fractal Web."

The computer graphics/chaos revolution has generated new kinds of artists. Britisher William Latham has become a sculptor who uses the computer screen instead of marble or clay.

Itsuo Sakane, a Japanese "science-art critic," describes Latham's work as eliciting "some kind of shock on being confronted with a strange and weird form that seems to have come into existence on a planet in some other galaxy and to have gone through an evolutionary process completely different from that followed on planet Earth. These forms seem to have been born of something both organic and inorganic . . . Yet there is also a sense of nostalgia, a sense of having before you the prototype of some form of life which you have seen somewhere long ago."

Latham uses fractal geometry as well as other computer graphics techniques in his work. He says, "In the past, artists such as Van Gogh, Cezanne, Monet have been concerned with representing the natural world, for example, Van Gogh's sunflowers or Monet's water lilies. What I am trying to do is produce my own version of the natural world. . . . The viewer is looking at a distorted synthetic nature, as though in a dream."

In this sculpture, which he calls "Inside Form," Latham covered the scaled spiral shape with a fractally patterned skin.

question of democratized art pointedly. Referring to the ability of people with simple algorithms and small computers to generate strange attractors and ornate designs of the Mandelbrot set, Pickover mused, "I'm wondering if it is disturbing to artists that a high school student can now produce these types of pictures which most of the people would call beautiful while they wouldn't necessarily care about 'true art.'"

Thus the question is, what is true art? Is it what is pretty, intriguing, made of forms that are both orderly and chaotic? If so, the Mandelbrot set images have these qualities. Are we approaching an era in which the fractal computer will replace the artist's intuition? While the answer is, "Probably not," the fractal qualities of self-similarity and simultaneous chaos and order do seem to be helping illuminate something important about the nature of art.

Consider the self-similarity of random fractals (like the fractal imitations of trees and mountains) and computer-generated nonlinear fractals (like the Mandelbrot set), where patterns at different sizes recur. Picture the warty gingerbread man of the Mandelbrot set who keeps reappearing like a magician's rabbit amidst permutations of the swirls, folds, and fireworks that stud the sky over the set's infinite coastline. Without a doubt the set is beautiful and variable, but perhaps after a while a little too predictable—of course, not literally predictable (it isn't literally predictable), but psychologically predictable. Perhaps it seems almost, after a time, a little boring. Now compare Mandelbrot art to universally acknowledged examples of "true art"—a Picasso or Brueghel or Shakespeare—those enduring works of any period, style, or culture that retain their vitality even after our repeated encounters with them. The great poem or painting is always new, always a mild surprise. Mona Lisa's smile, for example,

Parisian photographer Marie Bénédicte Hautem has captured the fine layering of self-similar detail that constitutes the Paris Opera Building. Mandelbrot himself cited the structure as an example of the scaling feature of the fractal geometry he invented. "One of my conclusions," wrote Mandelbrot, "is that it is fruitful to call Mies van der Rohe's buildings scalebound—a term a physicist would use to describe a flawless crystal and the solar system—and to call the Paris Opera House a scaling building—the term scaling also being applicable to typical views of the Alps and to the visual characteristics of many other objects in nature." As one walks down Rue de l'Opera, the closer one gets, the more of the building's self-similar detail comes into view. Mandelbrot's seemingly odd comparison of this baroque Beaux Arts building to objects in nature highlights the fact that though works of art may often look very different from "realistic" objects, the deep intent of many artists is to create forms that exhibit something of the inner structure and life to be found in nature's forms.

remains an enduring enigma. The chaologists who study the inner workings of the brain have come up with results that, by extrapolation, may suggest why we perceive great art as we do.

Brain scientists like Walter Freeman and Paul Rapp say that a healthy brain maintains a low level of chaos which from time to time self-organizes into a simpler order when presented with a familiar stimulus. In experiments done by Freeman and colleagues, a rabbit was given a familiar scent to sniff, and graphs of the pattern of electrical activity in the rabbit's olfactory bulb became simpler: The graph shifted from a strange attractor to a less-strange attractor. When the rabbit was given an unfamiliar scent, however, the normal strange attractor became even *stranger*. But this effect

Fascinated by the ideas of chaos and fractals, architect Peter Anders designed the interior of his loft apartment in the shape of a "strange attractor." Like the strange attractors plotted on computer screens by chaol-

continued from previous page

ogists, the visual lines wind through the space, creating a paradoxical sense of both infinitude and repetition, fragmentation and unity. This is the essence of the fractalist's aesthetic.

lasted only a while. Soon the unfamiliar scent became familiar, the rabbit's brain "habituated" to it, and the creature's brain graphs grew simpler. Since scientists believe that in a human brain similar processes occur, we might speculate that the form of an enduring work of art somehow resists the brain's tendency toward habituation. A great work seems to evoke a new, wild strange attractor every time the human brain encounters it. No matter how many times we read some great poem, listen to some great symphony, or gaze at some great painting, no matter how familiar we are with that work, it remains, at some important level of our perception, unfamiliar. The key is ambiguity created by artistic self-similarity.

This woodcut, "Waterfall in Yoshino," by the Japanese painter Hokusai (1760—1849), uses a scaling of reflectaphors traditional among Asiatic artists. The central reflectaphor here is subtle, but pervasive—a form that recurs in variations, providing a sense of unity, diversity, and wholeness to the work. This reflectaphoric form might be roughly described as a crab's claw. The claw shape appears at various scales and in numerous trans-forma-tions in the veg-etation, in the water, in the rocks (notice the claw at the root of the red rock at the bottom of the woodcut). The ochre horse forms the base of the claw of rock running along the river to the right. The horse also forms its own subtler claw-form with the arch of its neck. The two straining men make another subtle claw.

When painters juxtapose multiple self-similar forms and colors on canvas, or composers transform a sequence of notes into multiple self-similar forms by varying the rhythm and projecting the sequence of notes into different sections of the orchestra, they create a tension that gives birth to lucid ambiguities. Such artistic juxtapositions might be called "*reflectaphors*" because the self-similar forms reflect each other yet contain, like metaphors, a tension composed of similarities *and* differences between the terms. This reflectaphoric tension is so dynamic that it jars the brain into wonder, awe, perplexity, and a sense of unexpected truth or beauty.

To make great artworks, artists must find just the right distance between the terms of their reflectaphors, just the right balance of harmony and dissonance to create tension and the illuminating ambiguities that can flow from it. That proper balance is the one that catches the brain's processing by surprise and subverts habituation. It's the balance that forces our brains to experience the words or forms or melodies as if for the first time, every time, no matter how many times we have encounted them before. Artists find reflectaphoric harmony by testing the distance between the self-similar terms in their own brains first. A poet revising a poem may read over a line literally hundreds of times. Does the metaphor still have a jolt of surprise after all those readings? If so, it is a reflectaphor: a juxtaposition of terms that are both self-similar and different and as a result help open the mind.

So the fractals of the Mandelbrot set are almost art, but not quite. The parts are too similar, or in some cases too different from each other, to produce the kind of ambiguity-filled reflectaphoric webwork characteristic of a great work of art. Art is much more than a permutation of similar forms. It is creative in a way that is analogous to the creativity in nature: each form and gesture in an artwork has autonomy and yet its self-similarity draws it into an interaction with other forms and gestures in the piece to generate an environment that forces us to continuously realize the artwork is alive and dynamically in motion. Moreover, just as each single beetle or killer whale implies the whole of nature, Beethoven's symphony in its moods and rhythms implies the whole of everything, including ourselves.

In this classic landscape, "The Harvesters," by sixteenth-century Flemish painter Pieter Brueghel, the Elder, the artist creates reflectaphors by visually comparing, contrasting, and interweaving Euclidian forms. Note, for example, how many variations and how many scales there are of the triangular form of the haystack. It can be seen in the posture of the man lying under the tree, in the sharp perspective of the corridor of hay the man with the red jug is emerging from, in the house roofs. Note the circles: One is formed by the haystack and the two harvesters on the right. Another is formed by the line of mown hay joining the road on the left, joining the arc of green leading out to the water, and then curving around. A third circle is formed by the group of people eating in the right foreground. All these circles are incomplete and ragged; the triangles are peaked with a blobby circular form. Note the rectangular shapes; these are interwoven with the circles and the triangles. For example, the road on the left forms part of a rough rectangle as well as being part of the large rough circle that dominates the center of the painting. Note there is a vertex in the road that runs horizontally in the painting which suggests the peak of a triangle, and there are vertices all over the line of cut hay: These create more triangles. Notice how the road the haywagon is traveling on mirrors the line of mown hay in the foreground: it's both similar and different. Overall, Brueghel has found a way to interweave and "fractalize" Euclidian forms, creating a sense of simultaneous symmetry and asymmetry.

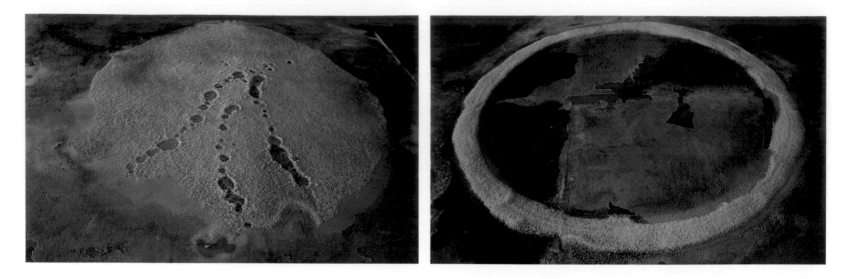

Could some formula or algorithm, some powerful but subtle feedback equation, one day enable us to churn out reflectaphors with the right distance, the right harmony of similarity and difference? Two Swiss scientists have devised a fractal algorithm where, as *The New York Times* put it, "mathematical extracts of the music of J. S. Bach could serve as matrices upon which new compositions of Bach-like music could be constructed, 'comparable in quality' to those of the composer himself." The premise of this approach is questionable, for if Bach's music is more than self-similar—if it is composed of reflectaphors—then it is highly doubtful that fully creative juxtapositions could be manufactured even by the richness of a fractal algorithm. Creating a network of musical reflectaphors (creating a great work of art) requires constant attention to the functioning of the human brain as it is listening to the composition, in order to find the harmony and dissonance between terms that will allow the strange attractors in the brains of artist and audience to resist habituation. It seems self-contradictory to think that a mechanical, if unpredictable, algorithm could accomplish this immensely subtle task. More likely the result will be a merely interesting, but ultimately lifeless, Bach imitation.

Artists are artists for their ability to make reflectaphors that capture their vision—that is, for their ability to project into a concrete form (painting, poem, music) their unique perspective on the whole (and each of us has a unique perspective on the whole, though we don't all make reflectaphors to express it). Each great work of art is a kind of microcosm or mirror of the universe. That means that each great artist's personal vision must also reflect the whole, which means reflecting the mysterious chaos and order of life itself.

The self-similarity of reflectaphors is much richer than the self-similarity in mathematical fractals, and allows each artist in each generation in each culture to develop a unique approach. The Flemish painter Brueghel created reflectaphors out of self-similar Euclidian forms repeated at different scales, trans-

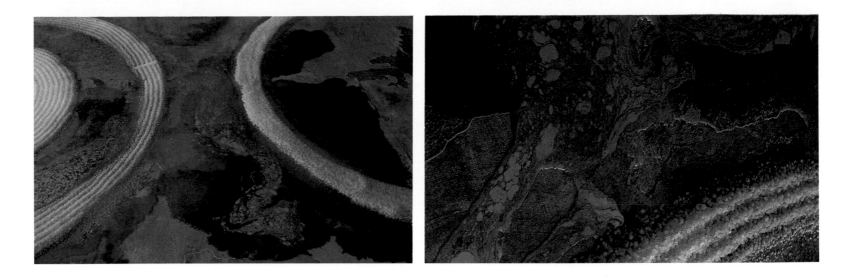

Eve Laramée uses copper, salt, and water to create her sculptures on site. She says, "My work has to do with the correspondence between nature, humankind, and self. It has to do with natural processes unfolding over time. I'm interested in removing the hand of the artist to a certain degree and just taking my chances. I set up the framework for a collaboration with nature. . . . Embodied within a sculptural form, my work creates environmental situations which isolate and record the change of matter from one state to another. My work brings forth the residue or traces left by the processes of evaporation, sedimentation, chemical reaction, and crystalization. It is always in a state of change"—that is, a state of evolving fractal patterns.

Laramée adds, "I see order and chaos as moments in a larger continuum. They are where you look for them; in art, they are what you choose to isolate or frame. I don't see order and chaos as separate entities, but as different types of sameness." She does not consider herself a "fractalist" or "fractalier," she says, "or do I consider my work 'fractal art.' I do not believe that such a movement or style exists. I believe my work dovetails or cross-references with, or is a 'reflectaphor' of, chaos and fractals. . . . I am skeptical of artists who try to create their own movement or label or terminology. I find it pretentious."

forming them into landscapes that are both rugged and regular, symmetrical and asymmetrical, active and frozen. Picasso and Braque created reflectaphors by breaking down objects into facets and then visually comparing these broken facets. The suprematist school of Russian painters, active about the time of the Communist Revolution, laid down large blocks of color on canvas, searching for a shape, size, and hue for the block so as to make it appear both simultaneously static and about to fly off the painting; the idea was that the same form should project diametrically opposing states.

Each generation of artists explores new ways to make reflectaphors. In some cultures the changes in reflectaphoric technique from one generation to the next amount to only nuances, as was the case with the subtle changes over

hundreds of years of Chinese landscape painters. In other cultures, such as our own, emphasis on the value of "originality" and the individuality of the artist spawns startling shifts in the way reflectaphors are made from one generation to the next. Consider, for example, the shift in methods from the impressionists to the cubists. The fractalist artists are no different (except, of course, that they are very different). Listen to fractalist painter Carlos Ginzburg talk about his aesthetic. He is responding to the Mandelbrot set images of German mathematician Heinz-Otto Peitgen, whose book *The Beauty of Fractals* has been widely praised for its gorgeous "artistic" images. Ginzburg insists that he and his fellow artists are looking for something different. "We want," he explains, "to present the beauty of 'kitsch' fractals against the 'Beauty of Fractals' of Dr. Peitgen. If we cannot go beyond Dr. Peitgen's positions, fractal art will be only a kind of 'scientific ready-made,' with really very little interest. The structural beauty of Peitgen's fractals is their 'good gestalt,' their inner harmony, their magnificent instability, the fact that they are new forms, a pure invention of Mandelbrot's genius. These kinds of fractal belong to the Renaissance; they offer the most traditional conception of beauty, maybe a modernist conception of beauty. What is certain today is the evidence that visual art got rid of this conception since the beginning of the twentieth century, or at least since 1960."

In making his manifesto for "kitsch" fractals, Ginzburg is being more than merely outrageous; he is doing what artists have always done in order to keep us alive to the mystery of life. He's making new reflectaphors and a new content for old reflectaphors. Thus art does not progress but tries in each generation to connect the unique spirit of a time with a primordial mysterious insight that lies deeper than chaos.

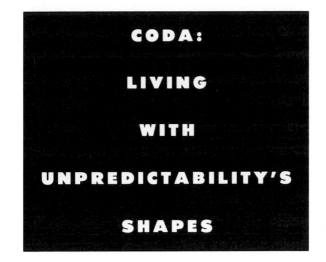

CODA:

LIVING

WITH

UNPREDICTABILITY'S

SHAPES

•

It is a constant idea of mine,

that behind the cotton wool [of

daily reality] is hidden a

pattern; that we—I mean all

human beings—are connected

with this; that the whole

world is a work of art; that we

are parts of the work of art.

        —Virginia Woolf in *A*

        Sketch of the Past.

•

Artists understand the ideas of fractals and chaos intuitively, and in their aesthetic response to the new science may lie its true importance. Whatever the study of fractals and chaos may bring in terms of practical applications, the deepest gift may be the opportunity these ideas offer for radically changing the way we look at nature. Fractals have the power to help us change our values in areas that may ultimately involve our survival on the planet. Aesthetics, which is about our sense of harmony in nature, has become a deadly serious activity.

The question is, shall we inhabit a world shaped (as we have long believed) by lifeless mechanically interacting fragments driven by mechanical laws and awaiting our reassembly and control? Or shall we inhabit a world—the one suggested by fractals and chaos—that is alive, creative, and diversified because its parts are unified, inseparable, and born of an unpredictability ultimately beyond our control?

The difference between these two world-views could not be more stark. As one commentator puts it, the thrall of the old mechanistic aesthetic has today led to a growing suspicion that order, at least as practiced by humankind, actually leads to disorder. From politics to science, humankind seems to be its own greatest threat. Every solution seems to generate its own problem. Planting orderly, genetically tailored trees to replace forests clear-cut by lumber companies leads to devastation of the new growth by pests and disease and the extinction of species. Damming up the Nile River to control floods and provide electric power depletes the soil downstream and increases water salinity. Jungian psychoanalyst John R. Van Eenwyk says, "So where chaos was once seen to undermine order, now order itself is also the culprit. Has science somehow slipped through the looking glass?"

Many scientists are immensely attracted to the new (and perhaps primordial) aesthetic described in this book. Art critic Klaus Ottmann speculates that the attraction results from the fact that scientists have long been starved of the joy that comes with focusing in an unfettered way on the visual dimensions of their work with nature. All of the scientists whose images are displayed here have clearly discovered that joy in their research with chaos and fractals. Several scientists, like neuroscientist Paul Rapp and Gottfried Mayer-Kress, have even been inspired to engage in collaborations with artists as a result of their research. In 1990 Mandelbrot teamed up with Pulitzer Prize–winning composer Charles

Wuorinen to create a multimedia performance at New York's Lincoln Center.

In his book *The Reenchantment of the World,* Morris Berman calls for an aesthetic that could turn our science (our knowledge of the world) into art. The aesthetic of fractals and chaos holds out that promise. But taking on such a challenge requires a sense of adventure and a certain courage. It means giving up absolute faith in our ability to control our environment ("science will save us"), and instead making our life out of unpredictability's shapes. It means attaining a certain humility about our place in the cosmos.

But there is also a serious danger that the concept of fractals and chaos will be transformed into a more sophisticated—even more totalitarian—version of our old mechanical view of life. Chaologists' ability to use simple formulas to generate complexity on the computer may soon convince them that humankind can in fact control complexity and master the dynamical forces of nature. In the past it was just this hubris that led us to a 500-year orgy of cutting nature down to fit our preconceived ideas—virtually simplifying our world out of existence. Social anthropologists say that in an earlier shift from oral to written cultures we learned to simplify reality in order to record it. Now, with the rise of the computer, we have an instrument (ironically, the very instrument that discovered fractals and chaos) that can digest such immense complexity we may be deluded into thinking we have recorded in its circuits the keys to creation.

This danger is real. In an October 1991 *Science Times* article describing an experiment that demonstrated that ecosystems are naturally subject to chaos, one researcher noted that the discovery of chaos upset the old idea that nature is "balanced." He concluded, "It really cuts the legs out from underneath this position that all we really have to do is leave these [ecological] systems alone and everything's going to be ducky. What we have to do is understand how these systems behave and *then we as people can decide what we want, how to manage them appropriately*" (emphasis added). At a vital level this scientist seems to have missed the point about chaos—at least the philosophical point. Fractals and chaos tell us about the inherent value of living in a world that springs beyond our control. Such a world enriches and invigorates our curiosity and awe, and that is why artists have responded intuitively to these ideas.

Perhaps at some level we will all need to become artists and chaologists in order to save it.

CREATING FRACTALS ON HOME COMPUTERS

The graphic capabilities of computers have forever changed the ways mathematicians and scientists do math and science. The once dry abstractions of nineteenth-century formulas can now be rendered into brilliant patterns and colors that tell the story of complex relationships and intricate feedback dynamics at a single glance.

Paradoxically, complex fractal graphics derive from relatively simple arithmetic. To make these pretty pictures in any detail, however, requires millions of tedious iterative operations—the perfect job for a microprocessor rather than a human being. Although sophisticated machines can do the job faster, home personal computers can create elegant fractal images as well.

Listed below are a number of off-the-shelf fractal imaging programs—organized by the kinds of machines on which they can be run. Also listed for serious do-it-yourselfers are a number of sources for generalized algorithms.

Be forewarned: The arithmetic may be simple for a computer, but calculation and graphic display time for zooming in on complex fractal objects like the Mandelbrot and Julia sets can progress extremely slowly on low-end machines. It helps to have a math coprocessor on board, and a newer generation of central microprocessor. Some programs are intended for monochrome display, some for 8-, 16-, or 32-bit color. *Be sure to check to see that the software matches your hardware configuration. Into the infinite depths, explore!*
—Doug Smith, May 1992

FRACTAL SOFTWARE FOR IBM

SHAREWARE

The following modestly priced programs, available from Public Brand Software (P.O. Box 51315, Indianapolis, IN 46251, 800-426-3475), allow exploration of Julia and Mandelbrot sets and creation of your own iterative graphic patterns: Mandelbrot 1–MA40.0; Mandelbrot 2–MA42.1; Mandelbrot for Windows–MA45.0.

COMMERCIAL PROGRAMS

The following are all available from Media Magic (P.O. Box 507, Nicasio, CA 94946, 800-882-8284).

Fractal Grafics, Cedar Software. A fractal geometry drawing program for creating naturalistic objects and abstract geometries via mouse and keyboard for IBM PC compatibles.

The Desktop Fractal Design System, Barnsley, Academic Press. Together with Barnsley's *Fractals Everywhere* text, this instructional software is a powerful primer for engineers, scientists, and other students of fractal geometry. On 5.25-inch disks only and requires EGA or VGA card, 640K RAM, DOS 2.0. (Available directly from Academic Press, 800-321-5068.)

Chaos, the Software, AutoDesk, Inc. A software package to accompany James Gleick's best-selling *Chaos, Making a New Science* (Viking). Includes a manual to explain the mathematics of a variety of strange attractors. Requires IBM PC/XT/AT.PS/2 or compatible with 640K RAM.

Fractal Creations, by Timothy Wegner and Mark Peterson. Bundled with a book of the same title this program (available on 5.25-inch disc only) allows the user to create fractals simulating natural objects—ferns, clowds, mountains—as well as to iterate any pattern of the user's own creation. It also generates stereoscopic fractals, which can be viewed with the 3-D glasses included in the package.

Fractools, Bourbaki, Inc. A sophisticated graphic program for the creation of beautiful fractal-art images. Many colorizing and "slide show" recording features. Comes on both 3.5-inch and 5.25-inch disks, thus requiring hard disk and IBM PC/XT/AT/PS2 or compatible with 512K RAM, DOS 2.0 and EGA or VGA.

Fraczooms, Bourbaki, Inc. Turns the computer into a high-speed fractal microscope for probing the recesses of a wide variety of functions including Lotka Volterra equations, trig functions, Newton's Method, Pickover's as

well as Mandelbrot and Julia sets and other attractors. The requirements are the same as for *Fractools*.
Pickover Sampler Software, Bourbaki, Inc. A hands-on amplification of Cliff Pickover's spectacular book *Computers, Pattern, Chaos, and Beauty* (St. Martin's Press), which allows the user to enter values in order to modify and experiment with various equations.

FRACTAL SOFTWARE FOR MACINTOSH

SHAREWARE

The following disks are available at modest member and nonmember prices through Mac Group, Boston Computer Society, 48 Grove Street, West Somerville, MA 02144, (617) 625-7080. Members may download individual programs via bulletin-board services. *SuperMandelzoom* is a black-and-white, very fast program with an elegant "microscope stage" front end allowing easy maneuvering in the fractal world. Target areas are displayed in a series of ever higher resolution displays, allowing repeated zooming or change of view without long waiting periods. Color B: *Ani-Mandel;* Color C: *Fractal, MandelColor;* Color E: *Mac II Fractal;* Color I: *Mandella;* Education B: *Fractal Contours, Mandelbrot, SuperMandelzoom;* Education C: *Fractals;* Education G: *Lifemaker;* Education E: *Cellular Automata;* Graphics E: *Mandelzot;* Graphics F: *Fractal Magic, More Chaos!*.

COMMERCIAL PROGRAMS

The following are available from Media Magic, P.O. Box 507, Nicasio, CA 94946, 800-882-8284.
The Beauty of Fractals Lab, Eberhardt and Parmet, Springer Verlag. A versatile and powerful software package that accompanies Pietgen and Richter's elegant books *The Beauty of Fractals, The Science of Fractal Images,* and *Fractals for the Classroom.* Sequential increases of resolution accelerate results. Includes intriguing 3-D topographic renderings of the fractal universe and color-editing features. Runs in 256 colors on all Mac II's.
Mandelmovie, Dynamic Software. Runs on all Macs and supports color on Mac II's, with panning and zooming features to probe Mandelbrot, Julia sets, and other attractors. Includes animation utility in order to create movie zooms, and a HyperCard tutorial. Can export and print as PICT files.
Fractasketch, Dynamic Software. A graphic arts and teaching program that will create iterative images from basic user-created shapes (library of images is included with the program). Includes file compression features.
Fractal Attraction, Lee and Cohen, Academic Press. A "draw it yourself" fractal program: Enter a freehand design in one window and its fractal transformation is rendered in another. Runs on Mac Plus and above, and in color on II's.
The Desktop Fractal Design System, Barnsley, Academic Press. Runs on Mac Plus or higher. See IBM version.

MISCELLANEOUS PUBLICATIONS

Amygdala, Box 219, San Cristobal, NM 87564. A newsletter devoted to fractal arcana and strange attractors.
Art Matrix, P.O. Box 880, Ithaca, NY 14851. A catalog of books, videos, posters, slides and, yes, T shirts.
Discovering Apple Logo: An Introduction to the Art and Pattern of Nature by David Thornbury (Reading, Mass.: Addison-Wesley, 1983). For owners of Apple IIe's and Commodore 64s running LOGO, this book describes numerous recursive graphics that can be used to create fractals found in the natural world using simple "turtle" commands.
Leonardo: Journal of the International Society for Arts, Sciences, and Technology, 672 South Van Ness, San Francisco, CA 94110. A quarterly publication that deals with matters at the boundary between the sciences and the arts, including chaos, fractals, and their relationship to visual arts and music.

PSEUDO CODE

Dewdney, A. K., "Computer Recreations," *Scientific American 253;* July 1987; February 1989.
Swaine, Michael, "Fooling Mother Nature with Fractal Flora," *MacUser,* March 1989, pp. 213–25.

CONTRIBUTERS' BIOGRAPHIES

Juan Acosta-Urquidi is a member of the Department of Ophthalmology at the University of Washington, Seattle, Washington.

American Museum of Natural History is located on Central Park West at 79th Street in New York City.

Peter Anders is an architect with Kiss Cathcart Anders in New York City.

Jenifer Bacon, a graphic artist in Irvine, California, has collaborated with Gottfried Mayer-Kress on computer images of chaos.

Otto Baitz is a photographer of architecture and interior design with offices in Red Bank, New Jersey.

Per Bak is a senior scientist with the Brookhaven National Laboratory, Department of Physics, Upton, New York.

Michael Barnsley is a physicist at the Georgia Institute of Technology and has founded his own company, Iterated Systems, Inc., in Norcross, Georgia.

Michael Batty is professor of geography at the State University of New York at Buffalo.

Edward Berko is a visual artist residing in New York City.

Pieter Brueghel, The Elder, was a Flemish painter, 1520?–1569.

Christopher Burke is a photographer with Quesada/Burke in New York City.

Scott Burns is an associate professor of general engineering at the University of Illinois at Urbana-Champaign.

Joe Cantrell teaches photography at the Pacific Northwest College of Art in Portland, Oregon.

Loren Carpenter is an animation scientist with Pixar in Richmond, California.

The Collection of Historical Scientific Instruments is located at the Science Center at Harvard University, Cambridge, Massachusetts.

Lilia Ibay de Guzman is a graduate student at the USDA/ARS, Honey-Bee Breeding, Genetics & Physiology Research in Baton Rouge, Louisiana.

Earth Observation Satellite Company, located in Lanham, Maryland, is responsible for operating US LANDSAT Remote Sensing Satellite and for marketing and selling LANDSAT data worldwide.

Fidia Research Laboratories is located in Abano Terme, Italy.

Mike Field is a professor of mathematics at Sydney University, Australia.

Deborah R. Fowler is a Ph.D. student at the University of Regina in Canada.

Walter J. Freeman is a member of the Department of Physiology-Anatomy at the University of California, Berkeley.

Norma Fuller is a Ph.D. student at the University of Regina in Canada.

Carlos Ginzburg is a fractalist painter residing in Paris.

Tiana Glenn is the video production specialist at the Boise Inter-Agency Fire Center.

Goddard Space Center is in Greenbelt, Maryland.

Ary Goldberger, M.D., is an associate professor of medicine at Harvard Medical School and director of electro-cardiography, co-director of the Arrhythmia Laboratory, Cardiovascular Division, at Beth Israel Hospital in Boston, Massachusetts.

Dr. Joseph H. Golden is senior meteorologist with the Office of the Chief Scientist at the National Oceanic and Atmospheric Administration in Washington, D.C.

Martin Golubitski is a professor in the Department of Mathematics at the University of Houston.

Celso Grebogi is a professor at the Laboratory for Plasma Research at the University of Maryland at College Park, and a member of the University of Maryland Chaos Group.

Owen Griffin is a researcher at the Naval Research Laboratory in Washington, D.C.

Margaret Grimes is a professor in the Art Department at Western Connecticut State University. She exhibits her paintings at the Blue Mountain Gallery in New York City.

David B. Grobecker is the scientific director of the GAIA Marine Institute in Kailua-Kona, Hawaii.

James Hanan is a doctoral student at the University of Regina in Canada.

M. B. Hautem is a photographer and painter residing in Paris.

Daryl Hepting is a master's student at the University of Regina in Canada.

John A. Hoffnagle is a physicist with IBM Research in San Jose, California.

Katsushika Hokusai (1760–1849) was a Japanese painter and print maker. He is considered one of the six great Ukiyoe masters and the founder of the school of landscape artists that dominated this form during its last phase.

Lawrence Hudetz is a photographer residing in Portland, Oregon.

Dr. Eugenia Kalnay is the chief, Development Division of the National Oceanic and Atmospheric Administration at the National Meteorological Center in Washington, D.C.

Nancy Knight is a researcher with the National Center for Atmospheric Research/National Science Foundation in Boulder, Colorado.

E. L. Koschmieder is on the faculty of the University of Texas at Austin's College of Engineering and Center for Statistics and Thermodynamics.

Kamala Krithivasan is a professor at the Indian Institute of Technology in Madras, India.

Robert Langridge is with the Computer Graphics Laboratory at the University of California.

Eve A. Laramée is an artist residing in New York.

William Latham is an artist sponsored by IBM at the UK Scientific Center in Winchester, Hampshire.

John Lewis is a computer graphics researcher with the NEC Research Institute in Princeton, New Jersey.

Aristide Lindenmayer was a professor and head of theoretical biology at the University of Utrecht in The Netherlands at the time of his death in 1989.

Edward Lorenz is a meteorologist at the Center for Meteorology and Physical Oceanography at the Massachusetts Institute of Technology.

David Malin is with the Anglo-Australian Observatory, Epping Laboratory, in Epping, Australia.

Mario Markus is a physicist with the Max Planck Institute in Dortmund, Germany.

Gottfried Mayer-Kress is researching nonlinear dynamics at the Sante Fe Institute.

William A. McWorter, Jr., is a professor in the Mathematics Department at Ohio State University.

Paul Meakin is a researcher with the Central Research & Development Department Experimental Station at E. I. DuPont de Nemours & Company, Inc., Wilmington, Delaware.

Mark Meier is with the Institute of Arctic and Alpine Research at the University of Colorado at Boulder.

Nachumae Miller is a painter residing in New York City.

Mark Moore is with the Northwest Avalanche Center in Seattle, Washington.

Steven D. Myers is with the Mesoscale Air-Sea Interaction Group at Florida State University in Tallahassee.

National Aeronautics and Space Administration launched *Voyager 1* which encountered Jupiter and Saturn in 1979 and 1980, *Voyager 2* which encountered Jupiter in 1979, Saturn in 1981, and Uranus in 1986.

National Cancer Institute is located in Bethesda, Maryland.

National Optical Astronomy Observatories is located in Tucson, Arizona.

National Severe Storms Laboratory is located in Norman, Oklahoma.

Michael Norman is an astrophysicist at Los Alamos National Laboratory. He has used the supercomputer at the University of Illinois at Urbana-Champaign to model the behavior of interstellar jets.

Office National D'Études et de Recherches Aérospatiales is located in Chatillon, France.

Peter Oppenheimer is with the New York Institute of Technology, Computer Graphics Lab.

Clifford A. Pickover is a staff member at the IBM Thomas J. Watson Research Center in Yorktown Heights, New York.

David Plummer is with the National Meteorological Center in Washington, D.C.

Przemyslaw Prusinkiewicz is a professor in the Department of Computer Science at the University of Calgary in Alberta, Canada.

Bill Quinell is a member of the Art Department at Western Connecticut State University.

P. E. Rapp is a professor in the Department of Physiology at The Medical College of Pennsylvania.

Rollo Silver publishes a newsletter, *Amygdala* (devoted to fractals and the Mandelbrot set), from San Cristobal, New Mexico.

Peter Siver is a member of the Department of Botany at Connecticut College.

Doug Smith develops interactive science exhibits, using multimedia technology, for science museums across the country. He lives in Boston.

Homer Smith is co-founder of Art Matrix, located in Ithaca, New York.

Allan Snider is a master's student at the University of Regina in Canada.

Joel Sommeria is a research physicist at École Normale Superiure de Lyon in France.

Harry Swinney is a professor of physics at the University of Texas's Center for Non-Linear Dynamics in Austin.

Lucinda Tavernise is a freelance illustrator living in Granville, Massachusetts.

G. J. F. van Heijst is a professor at the Institute of Meteorology and Oceanography, University of Utrecht, The Netherlands.

Manuel G. Velarde is a professor of physics at the Autonomous University of Madrid.

Andreas Vesalias (1514–1564) was a Belgian anatomist, considered the founder of modern anatomy. His major work was "De humani corporis fabrica libri septem."

Britony Wells is a photography student at Western Connecticut State University.

Edward Weston (1886–1957) was an American photographer who gave more than seventy-five one-man shows and was the author of several photography books. His work is archived at the Center for Creative Photography in Tuscon, Arizona.

Arthur Winfree is a member of the Department of Ecology & Evolutionary Biology at the University of Arizona.

Jack Wisdom is a professor of physics at the Massachusetts Institute of Technology.

Lewis R. Wolberg, M.D., was a well-known psychiatrist practicing in New York City, and author of *Micro-Art: Art Images in a Hidden World.* He died in 1988.

Jerome J. Wolken is a professor in the Department of Biological Sciences at Carnegie Mellon University.

SUGGESTED READING

Books on Fractals, Chaos, and the Aesthetics of Self-Similarity for Readers with a Nonscientific Background

Briggs, John, and F. David Peat. *Turbulent Mirror: An Illustrated Guide to Chaos Theory and the Science of Wholeness.* New York: Harper Collins, 1989.

Garcia, Linda. *The Fractal Explorer.* Santa Cruz, Calif.: Dynamic Press, 1991.

Gleick, James. *Chaos: Making a New Science.* New York: Viking, 1987.

McGuire, Michael. *An Eye for Fractals: A Graphic & Photographic Essay.* Redwood City, Calif.: Addison-Wesley, 1991.

Peitgen, H.-O., and P. H. Richter. *The Beauty of Fractals: Images of Complex Dynamical Systems.* Berlin: Springer-Verlag, 1986.

Pickover, Clifford A. *Computers, Pattern, Chaos and Beauty: Graphics from an Unseen World.* New York: St. Martin's Press, 1990.

Prigogine, Ilya, and Isabelle Stengers. *Order Out of Chaos: Man's New Dialogue with Nature.* New York: Bantam, 1984.

Stewart, Ian. *Does God Play Dice: The Mathematics of Chaos.* Cambridge, Mass.: Basil Blackwell, 1990.

Image Credits

Page 13: Photo by Joe Cantrell.

Page 14: Peter A. Siver, *The Biology of Mallomonas: Morphology, Taxonomy and Ecology* (Kluwer Academic Publishers).

Page 17: National Optical Astronomy Observatories.

Page 20: Paul Meakin.

Pages 22–23: John Briggs.

Page 26: National Aeronautics Space Administration, Goddard Space Center.

Page 29: Nachumae Miller.

Page 31: P. E. Rapp.

Page 33: Reproduced by permission of Earth Observation Satellite Company, Lanham, Maryland, U.S.A.

Page 36: Copyright © 1992 Lawrence Hudetz. All rights reserved.

Page 37, top: Photo by Lewis R. Wolberg, M.D., *Micro Art* (New York: Harry N. Abrams, Inc.).

Page 37, bottom: Juan Acosta-Urquidi.

Page 38: David B. Grobecker, GAIA Marine Institute, Kona, Hawaii.

Page 39, top: Photo by Bill Quinell and Brittony Wells.

Page 39, bottom left: Photo by Lewis R. Wolberg, M.D., *Micro Art* (New York: Harry N. Abrams, Inc.).

Page 39, bottom right: Michael Barnsley, *Fractals Everywhere* (San Diego: Academic Press, 1988).

Page 40: Copyright © 1992 Lawrence Hudetz. All rights reserved.

Page 41: Lilia Ibay de Guzman, USDS/ARS, Honey Bee Breeding Genetics and Physiology Lab. 1157 Ben Hur Rd, Baton Rouge, LA 70820.

Page 44: Jennifer Bacon, from map by Gottfried Meyer-Kress.

Page 46: Per Bak.

Page 47, left: Photo by U.S. Geological Survey (Mark Meier).

Page 47, right: Mark Moore, Northwest Avalanche Center.

Page 50: Photo by David Malin, copyright © Ango-Australian Telescope Board/ROE.

Page 51, top: Courtesy of the Collection of Historical Scientific Instruments.

Page 51, bottom: National Aeronautics Space Administration.

Page 52: Jack Wisdom.

Page 53: National Aeronautics Space Administration.

Page 54: National Aeronautics Space Administration.

Page 56, left: National Aeronautics Space Administration.

Page 56, right: Photo by the National Severe Storms Laboratory, National Oceanic and Atmospheric Administration.

Page 57, top: Photo by the National Severe Storms Laboratory, National Oceanic and Atmospheric Administration.

Page 57, bottom: Curves produced by Edward N. Lorenz.

Page 59: Nancy Knight, National Center for Atmospheric Research, National Science Foundation.

Page 60: David Plummer and Eugenia Kalnay, National Meteorological Center, Development Division.

Page 62: Reproduced by permission of Earth Observation Satellite Company, Lanham, Maryland, U.S.A.

Page 63: Copyright © 1992 Lawrence Hudetz. All rights reserved.

Page 64 and page 65, top: William McWorter.

Page 65, bottom: Przemyslaw Prusinkiewicz and Kamala Krithivasan, 1988.

Page 66: William McWorter.

Page 67, bottom: Photo by Joe Cantrell.

Page 68, bottom: Daryl Hepting and Allan Snider, 1990.

Page 69: Tiana Glenn, Boise Interagency Fire Center.

Page 70: Photos by Bill Quinell.

Page 74: C. Pickover, *Computers and the Imagination* (New York: St. Martin's Press, 1991); and C. Pickover, *Computers, Pattern, Chaos, and Beauty* (New York: St. Martin's Press, 1990). All rights reserved.

Page 76: Homer Smith, Art Matrix.

Pages 77–80, bottom: Homer Smith, Art Matrix.

Page 79, top: Rollo Silver, Amygdala, Box 219, San Cristobal, NM, 87564.

Page 81: Homer Smith, Art Matrix.

Page 84: Loren Carpenter.

Page 85: Photos from Michael Batty, "Fractals: Geometry between dimensions," *New Scientist,* vol. 105, no. 1,450 (April 4, 1985): 31–35 (photos on pages 34 and 35); and Michael Batty, *Microcomputer Graphics: Art Design and Creative Modelling,* (London: Chapman & Hall, Co., 1987).

Page 86, top: Przemyslaw Prusinkiewicz and Aristid Lindenmayer, 1987.

Page 86, bottom: Michael Barnsley, *Fractals Everywhere* (San Diego: Academic Press, 1988).

Page 87: Przemyslaw Prusinkiewicz, 1986.

Page 88, right: Prusinkiewicz and Norma Fuller, 1990.

Page 88, left: Daryl Hepting and Przemyslaw Prusinkiewicz, 1990.

Page 91: Peter Oppenheimer, New York Institute of Technology, Computer Graphics Lab; based on Raspberry symmetry model by Haresh Lalvani.

Page 92: John Lewis.

Page 94: Reproduced with permission of Mike Field (Sydney, Australia) and Martin Golubitsky (Houston, Texas). These pictures arose out of a study of the effects of symmetry on chaotic dynamics.

Page 95, left: Peter Oppenheimer, New York Institute of Technology, Computer Graphics Lab.

Page 95, right: Neg. no. 2A 5144; courtesy Department of Library Services, American Museum of Natural History.

Page 96, top: E. L. Koschmieder, *Advances in Chemical Physics* (New York: John Wiley & Sons, 1974).

Page 96, bottom: Pictures made and supplied by Professor M. G. Velarde (Spain). The pictures appeared in Milton Van Dyke, *An Album of Fluid Motion* (Stanford: Parabolic Press, 1982); and in M. G. Velarde and Christiane Normand, "Convection," *Scientific American,* July 1980.

Page 97: Robert Langridge, Computer Graphics Laboratory, University of California, San Francisco, California.

Page 100: Copyright © 1992 Lawrence Hudetz. All rights reserved.

Page 101, top: Reproduced by permission of Earth Observation Satellite Company, Lanham, Maryland, U.S.A.

Page 101, bottom: Reproduced by permission of Earth Observation Satellite Company, Lanham, Maryland, U.S.A.

Page 102, top: Copyright © 1992 Lawrence Hudetz. All rights reserved.

Page 102, bottom: Copyright © 1992 Lawrence Hudetz. All rights reserved.

Page 103: Photo by Joe Cantrell.

Page 104, top: Copyright © 1992 Lawrence Hudetz. All rights reserved.

Pages 104–105, bottom: Copyright © 1992 Lawrence Hudetz. All rights reserved.

Page 105, top: "Tracks in Sand, North Coast, 1937"; photograph by Edward Weston; copyright © 1981 Center for Creative Photography, Arizona Board of Regents.

Page 106: Copyright © 1992 Lawrence Hudetz. All rights reserved.

Page 108: 1978 photograph from the laboratory of A. T. Winfree.

Page 109: Jerome J. Wolken.

Page 110: National Optical Astronomy Observatories.

Page 111, top: Mario Markus, Max-Planck Institute, Dortmund, Germany.

Page 111, bottom: Dr. Joseph H. Golden, National Oceanic and Atmospheric Administration.

Page 113, top: John Briggs.

Page 113, bottom: National Aeronautics Space Administration.

Page 114: J. Sommeria, S. Myers, and H. L. Swinney, University of Texas.

Pages 116–117: John Briggs, from video by Hugh McCarney.

Page 118: National Aeronautics Space Administration.

Page 119: Photo by Joe Cantrell.

Page 120: C. Pickover, *Computers and the Imagination* (New York: St. Martin's Press, 1991); and C. Pickover, *Computers, Pattern, Chaos, and Beauty* (New York: St. Martin's Press, 1990). All rights reserved.

Page 121: C. Pickover, *Computers and the Imagination* (New York: St. Martin's Press, 1991); and C. Pickover, *Computers, Pattern, Chaos, and Beauty* (New York: St. Martin's Press, 1990). All rights reserved.

Page 124: A. Vesalias (1514–1564).

Page 125, top: National Cancer Institute.

Page 125, bottom: Fidia Research Laboratories.

Page 126: Dr. Ary L. Goldberger.

Page 127: Christopher Burke, Quesada/Burke, New York.

Page 128: With permission from Walter Freeman and *Scientific American*.

Page 129, top: C. Pickover, *Computers and the Imagination* (New York: St. Martin's Press, 1991); and C. Pickover, *Computers, Pattern, Chaos, and Beauty* (New York: St. Martin's Press, 1990). All rights reserved.

Page 129, bottom: C. Pickover, *Computers and the Imagination* (New York: St. Martin's Press, 1991); and C. Pickover, *Computers, Pattern, Chaos, and Beauty* (New York: St. Martin's Press, 1990). All rights reserved.

Page 132: Michael Norman, University of Illinois.

Page 133: NASA photo courtesy of Owen M. Griffin, Naval Research Laboratory, Washington, D.C.

Pages 134–135, top: Photo made by G. J. F. van Heijst and J. B. Flór (University of Utrecht, The Netherlands).

Pages 134–135, bottom: Office National D'Études et de Recherches Aérospatiales.

Page 136: Copyright © 1992 Lawrence Hudetz. All rights reserved.

Page 139: Jerrold E. Marsden, *Foundation of Mechanics* (Reading, Mass.: Addison-Wesley, 1978). Illustration was redrawn by Lucinda Tavernise.

Page 140: Michael Barnsley, *Fractals Everywhere* (San Diego: Academic Press, 1988).

Page 141, top: Photo by Joe Cantrell.

Page 141, bottom: Homer Smith, Art Matrix.

Page 142, top: C. Pickover, *Computers and the Imagination* (New York: St. Martin's Press, 1991); and C. Pickover, *Computers, Pattern, Chaos, and Beauty* (New York: St. Martin's Press, 1990). All rights reserved.

Page 142, bottom: Celso Grebogi, University of Maryland Chaos Group.

Page 143: Homer Smith, Art Matrix.

Page 144, left: Homer Smith, Art Matrix.

Page 144, right: C. Pickover, *Computers and the Imagination* (New York: St. Martin's Press, 1991); and C. Pickover, *Computers, Pattern, Chaos, and Beauty* (New York: St. Martin's Press, 1990). All rights reserved.

Page 145: John A. Hoffnagle.

Page 146: Scott A. Burns.

Pages 150–151: C. Pickover, *Computers and the Imagination* (New York: St. Martin's Press, 1991); and C. Pickover, *Computers, Pattern, Chaos, and Beauty* (New York: St. Martin's Press, 1990). All rights reserved.

Page 153: Mario Markus, Max-Planck Institute, Dortmund, Germany.

Page 154: Mario Markus, Max-Planck Institute, Dortmund, Germany.

Page 155: Mario Markus, Max-Planck Institute, Dortmund, Germany.

Page 159: Copyright © 1992 Lawrence Hudetz. All rights reserved.

Page 160: Copyright © 1992 Lawrence Hudetz. All rights reserved.

Page 163: Margaret Grimes.

Page 167: Carlos Ginzberg.

Page 168: Edward Berko, oil on wood, 48″ × 36″ (122cm × 91 cm), copyright © 1991. Courtesy: private collection, New York City.

Page 169: A still from *The Conquest of Form*, created by William Latham. Produced at the IBM UK Scientific Centre, Winchester, UK.

Page 170: M. B. Hautem.

Pages 171–172: Architecture: Peter Anders. Photos: Otto Baitz.

Page 173: Hokusai, courtesy of John Briggs.

Page 175: Pieter Breughel, The Elder; "The Harvesters," Metropolitan Museum of Art, New York.

Pages 176–177: Eve A. Laramée.

I N D E X

Page numbers in *italics* refer to illustrations.

Africa, *33, 101*
Ala River, 33
algae, *14*
Aliens, 92
Anders, Peter, *171–72*
animals, 36–41, *37, 38, 40, 41*
  birds, 115, 119–20
  evolution of, 37–41, 117, 120
  feedback in, *89*
  fish, *37, 40, 119*
  sense of smell of, 171–72
  see also insects
animal tracks, *105*
antimatter and matter, *88*
Apollo, 16
Apollo 11, 118
architecture, *65, 170, 171–72*
Argo Merchant, 133
art, 27–31, 158–64, 166–78, 180
  biometric, 129
  democratization of, 156, 169–70
  from fractal equations, *44,* 148–
    156, 169, 170, 174, 178
  holism in, 148–49
  irregularity in, 158
  painting, 27–28, *29,* 158, 164, *167,*
    168, 173, 174, 175, 176–77
  reflectaphors in, *173,* 174, *175,*
    176, *177*–78
  science and, 32–33, *44,* 80, 180–
    181
  sculpture, 32–33, 158, *169, 177*
  self-similarity in, 30, *88–89,* 148–
    149, *168, 172, 174*
  strange attractors and, 166, 172
  symmetry-chaos hybrids in, 94
  turbulence in, *136*
  see also photography
Art Matrix, 80, *81,* 123
asteroid belt, 52, *54,* 138
astronomy, 49–54
  beyond our solar system, *17, 50,*
    110, 132
  of our solar system, *47,* 49–54,
    51, 53, 54, 138–39
autocatalytic processes, 109–12, *109,*
  110
  see also feedback
avalanches, *47*

Babylonian mythology, 16
Babyloyantz, Agnes, 126–27
Bachelard, Gaston, 73
Bacon, Jenifer, *44*
badlands, *104*

Bak, Per, *46*
Bangladesh, *101*
barium ions, 143
"Barn Owls" (Siver), *14*
Barnsley, Michael, *39,* 86–87
Beauty of Fractal, The (Peitgen),
  178, 183
bees, *41*
beetles, 37
Belousov, Boris, 110
Belousov-Zhabotinskii (BZ) chemi-
    cal reaction, *108,* 109–10, *109,*
    111, 141
Berko, Edward, *168*
Berman, Morris, 181
Between the Acts (Woolf), 10
bifurcation points, 112
biomorphs, 120
birds, 115
  flight patterns of, 119–20
Blind Watchmaker, The (Dawkins),
  115, 117
Book of Changes, 147
Boston Globe, 36
brain, *125, 128, 171–72*
  EEG images of, 31–32, *31*
Braque, Georges, 177
British Association for the Advance-
    ment of Science, 133
British Columbia, *47*
Brookhaven National Laboratory, *46*
Brueghel, Pieter, the Elder, *175,*
  176–77
Brussels, Free University of, 126
Burns, Scott, 32, 149–50
butterfly mask of unpredictability,
  143

Calgary, University of, 86, 87
Cameroon, *33*
cancer cells, *125*
Cantor, Georg, 67
Cantor dust, 67
Cantor set, 66
Cantrell, Joseph, *13,* 28, 102–3, *103*
carp, *119*
Carpenter, Loren, *84*
cauliflower, *70*
Center for Non-Linear Dynamics,
    University of Texas, *114*
Cézanne, Paul, *169*
Chad, Lake, *101*
chaos:
  coinage of term, 12
  debate over definition of, 21

  see also specific topics
"Chaos Fractal 1985–86" (Ginz-
    burg), *167*
chemical reactions, *108,* 109–10,
  109, 111, 141
Chemung River, *62*
Chen, Kan, *46*
Chinese painting, 166, 178
Chinese philosophy, 16, *142*
Christianity, 16
circulatory system, *127*
climate, *see* weather
clouds, 118
  strange attractor of, *104–5*
coastlines, infinite, 62–63, *63,* 70
Columbia Gorge, *36*
Computer Graphics Lab, New York
    Institute of Technology, 90
computers, home, creating fractals
    on, 182–83
*Computers, Pattern, Chaos and
    Beauty* (Pickover), 154, 183
Computers and the Imagination
    (Pickover), 154
Conrad, Joseph, 131
Cornell University, 80
creative process, 28, 37–38
Creutz, Michael, *46*
cucumbers, *39*
cyclones, 112

Darwin, Charles, 35, 37–38, *39,* 108,
  115
Dawkins, Richard, 115, 117, 120
Descartes, René, 138
Dionysius, 16
DNA, 19, *97*
*Does God Play Dice: The Mathemat-
    ics of Chaos* (Stewart), 43
Dorn, Alfred, 93

Earth Observation Satellite Com-
    pany (EOSAT), *62, 101*
earthquakes, 112
electrocardiograms (ECGs), 126
electroencephalograms (EEGs), 31–
  32, *31*
electromagnetic fields, *142*
Emerson, Ralph Waldo, 37
entropy, 17, 108
epileptic seizures, 110, 126–27
Escher, Maritus, 166
Euclidian geometry, 24–25, *57, 62,*
  64, 90

  see also specific topics
"Chaos Fractal 1985–86" (Ginz-
    burg), *167*
art and, *175,* 176–77
  idealization in, 158
  photography and, 161
Europa, *51*
European Centre for Medium Range
    Weather, 59
Eustis, Mark, *62, 101*
evolution, 37–41, 117, 120

feedback, *89,* 109–12, *109, 110,* 112,
  176
  of chemical processes, 109–10
  in human body, *125,* 127–28
  in mathematics, *see* iterated and
    nonlinear equations
  negative vs. positive, 116–21, *117,*
    118, 119, 121
  sensitivity and, 19–20
  in snowflakes, *95*
  of sound, 107
  strange attractors and, 139–40
  video, 116, *117*
Fidia Research Laboratory, *125*
Field, Mike, *95*
fireflies, *109*
fish, *37, 40, 119*
Flash Art, 73
flies, flight of, *121*
folding processes, 133, *134, 142*
  in weather, 60, *143*
food supplies, *144–45*
Ford, Joseph, 10
forest fires, *69*
fractal equations, *see* iterated and
    nonlinear equations
Fractal Geometry of Nature, The
    (Mandelbrot), 71
fractals:
  coinage of term, 22, 61, 66
  definition of, 22–23, 71
  see also specific topics
"Fractal Web" (Berko), *168*
Freeman, Walter, *128, 171–72*
frogfish, *38*

gagonsa, 16–17
Ganges, *101*
geology, *67, 100,* 112, *141*
  see also landscapes, fractal
Georgia Institute of Technology, 86
Georgia Tech University, 10
ginger roots, *39*
Ginzburg, Carlos, *167, 178*
glaciers, *47*
Gödel's theorem, 27

Goldberger, Ary, 126–29
Golubitsky, Martin, 95
Gothic cathedrals, 166
gravity, 51
Greek mythology, 16
Grimes, Margaret, 28, 31, 158
Gruber, Howard, 37–38

Hardy, G. H., 147
Harvard Medical School, 126
"Harvesters, The" (Brueghel), 175
Hautem, Marie Bénédicte, 170
heart, human, 110, 126, 129
Heraclitus, 18
Hilbert curve, 64–66, 65
Hokusai, 173
holism, 21, 30, 80
    in art, 148–49
    definition of, 24
    turbulence and, 134
    in weather, 140
Houston, University of, 95
Hubbard, John, 80
Hudetz, Lawrence, 101–2, 104–5,
    136, 161, 164
human body, 71, 123–29, 125, 127,
    128
    background chaos of, 110–12,
        126–27, 171–72
    scaling in, 120
    strange attractors of, 31–32, 31,
        126, 128
    traditional image of, 124
hybrids of symmetry and chaos, 93–
    98, 95, 97
Hyperion, 52

IBM, 22, 46, 74, 84, 121, 129, 142,
    143, 154, 166
IBM-compatible computers, creating
    fractals on, 182–83
icebergs, 47
I Ching, 147
Illinois, University of, at Urbana-
    Champaign, 32, 132, 149
imitations of nature, fractal, 84–92,
    84, 87–89, 90–91, 92, 95, 97
immune system, 126–27
India, 16, 65, 139
insects, 37, 41, 109
    flight of, 121
    populations of, 144–45
"Inside Form" (Latham), 169
Institute of Meteorology and Ocean-
    ography, University of Utrecht,
    135
International Satellite Cloud Clima-
    tology Project, NASA, 59
ions, motion of, 143–45
Ireland, 113
Iroquois mythology, 16
iterated and nonlinear equations:
    art from, 44, 65, 148–56, 169, 170,
        174, 178

celestial mechanics and, 51–54
chemical reactions and, 108, 111
computers in solution of, 25–27,
    47, 135
history of, 45–47, 62–72
modeling of nature with, 25–27,
    45–47, 120–21, 121, 132, 133–
    135
sensitivity of, 148
turbulence and, 133–35
in visual imitations of nature, 85–
    90
see also Mandelbrot set

jellyfish, 37
jet streams, galactic, 132
Julia set, 154–56
    creating on home computers, 182,
        183
Jupiter, 51, 138
    giant eye of, 50, 53, 54, 112, 114

Keats, John, 27
Knight, Nancy, 59
Koch, Helge von, 64
Koch island, 66–67, 68, 70
Kolam, 65
Kolmogorov, Andrei, 138
Kolmogorov theorem, 138

lakes, 101
landscapes, classical, 175
landscapes, fractal, 13–14, 36, 37,
    47, 53, 99–106, 100, 101, 102,
    103, 104, 105, 106
    in art, 29, 164
    infinite coastlines, 62–63, 63, 70
    on Jupiter's moon, 51
    photography of, 36, 37, 47, 100,
        101–6, 101, 102, 103, 104, 105,
        106, 164
    satellite images of, 33, 101
    scaling in, 104
Laramée, Eve, 32–33, 177
Large Magellanic Cloud, 17
Late Night Thoughts on Listening to
    Mahler's Night Symphony
    (Thomas), 165
Latham, William, 169
leaves, 13, 37
    fractal imitations of, 86–87
Lewis, John, 92
lightning, 57
light solitons, 112
linearity, 19, 45–47, 51
    see also iterated and nonlinear
        equations
Lorenz, Edward, 15–17, 18, 56–59,
    143
Lorenz strange attractor, 59

Macintosh computers, creating frac-
    tals on, 183

Madrid, Autonomous University of,
    97
Mandelbrot, Benoit, 25, 68, 70, 71,
    75, 83, 84, 84, 90, 157, 170,
    180–81
    "fractal" coined by, 22, 61, 66
Mandelbrot set, 25, 26, 39, 40, 68–
    69, 70–71, 74, 74–81, 77, 79,
    81, 154
    as art, 169, 170, 174, 178
    creating on home computers, 182,
        183
    naming of, 80
    popularity of, 80–81
    weather and, 56
"Mandelbrot Stalks" (Pickover), 74
maple leaves, 86–87
Marduk, 16
Markus, Mario, 32, 151–52
Mars, 138
Martin, Benjamin, 51
Maryland, University of, 10
Massachusetts Institute of Technol-
    ogy, 15, 49, 52, 56, 92
mathematics, see iterated and non-
    linear equations; Mandelbrot
    set
matter and antimatter, 88
Mayer-Kress, Gottfried, 44, 180
Media Lab, Massachusetts Institute
    of Technology, 92
Medical College of Pennsylvania, 31
Meyers, Steven, 114
Michelangelo, 158
Mico-Art, Art Images in a Hidden
    World (Wolberg), 37
Miller, Nachume, 28, 29, 30
mold, slime, 110
Mondrian, Pieter, 158
Monet, Claude, 169
Moscow State University, 110
moss, 106
moths, gypsy, populations of, 144–45
mountains, 47
    fractal imitations of, 84, 85
Mt. Hood, 104–5
Mt. Rainier, 102
Mt. St. Helens, 100
movies, fractal techniques in, 84–85,
    84, 92
Museum of Modern Art, 29
music, 65, 176, 180–81
Mycelis muralis, 86
mythology, chaos in, 16–17, 18

National Aeronautics and Space
    Administration (NASA), 26, 59
Nature (Emerson), 37
New Scientist, 107
Newton, Isaac, 39, 45, 51, 138
Newton's method, 81, 138, 149–50,
    182
New York, 62

New York Institute of Technology,
    90
New York Times, 176
Nigeria, 33
nonlinear equations, see iterated
    and nonlinear equations
Norman, Michael, 132

oceans:
    oil spills in, 133
    satellite image of, 113
    waves in, 112, 113
Office National d'Études de Re-
    cherches Aérospatiales, 134
oil spills, 133
olfactory bulb, 128, 171–72
"On the Nature of Fractalization"
    (Berko), 168
Oppenheimer, Peter, 90–91
"Orchid, The" (Smith), 81
Oregon, 36, 103, 104, 136
Origin of Species (Darwin), 38
Orion Nebula, 17, 50
orreries, 51, 51
Ottmann, Klaus, 73, 145, 166, 180
outer space, see astronomy

painting, 27–28, 29, 158, 164, 167,
    168, 173, 174, 175, 176–77
paper, crumpled, 71
Paris Opera House, 166, 170
Parkinson's disease, 127
peacocks, 115
Peano, Giuseppe, 64
Peitgen, 76
Peitgen, Heinz-Otto, 76, 80, 178, 183
Pennsylvania, 62
Pensées (Woolf), 99
photography, 13, 28, 136, 161, 164
    computer images compared with,
        151–52
    of fractal landscapes, 36, 37, 47,
        100, 101–6, 101, 102, 103, 104,
        105, 106, 164
    of organic fractals, 23, 37, 38, 40,
        41, 70
Picasso, Pablo, 177
Pickover, Clifford, 74, 120, 121, 129,
    142, 154–56, 166–70, 182–83
Max Planck Institute, 32, 151
planetary motion, 47, 49–54, 138–39
plants, 23, 39–40, 39, 70, 106
    fractal imitations of, 85–87, 87–
        89
    trees, 69, 85, 103
Pluto, 49, 53
Poetics of Space, The (Bachelard),
    73
poetry, 174
Poincaré, Henri, 47, 51–52, 55
polio virus, 125
Pollock, Jackson, 166
polystyrene, 20

"Portrait of a 'Strange Attractor'"
(Hudetz), *104–5*
Prigogine, Ilya, 108
Princeton University, *90*
Prusinkiewicz, Przemyslaw, *65,* 86,
87–88

quantum mechanics, 27

rabbits, 171–72
Rapp, Paul, 31–32, *31,* 171–72, 180
"Raspberry, Garden at Kyoto" (Op-
penheimer), *91*
Raymo, Chet, 36–37
Reenchantment of the World, The
(Berman), 181
reflectaphors, *173,* 174, *175,* 176,
177–78
Regina, University of, 68
Rig Veda, 139
rivers, 33, *101*
Rösseler strange attractor, *108, 141*
Ruelle, David, 137
Russell, Bertrand, 148
Russell, John Scott, 112–13

Sakane, Itsuo, *169*
sandpiles, *46*
sandstone, 67
Santa Fe Institute, *44*
satellite images, *33, 101*
of oceans, *113*
Saturn, 52
rings of, *54*
scaling (worlds within worlds), *14,*
20, 23–25, *41,* 68, 134
in architecture, *170*
in art, 28, *173,* 176–77
in electromagnetic fields, *142*
in fractal imitations of nature, 85
in fractal landscapes, *104*
in Julia set, 156
Mandelbrot set and, *26,* 77–78
in organisms, 23, *23, 106,* 120
in weather, 23–24, 59–60
science, 27
art and, 32–33, *44,* 80, 180–81
chaos recognized by, 15
logic of nature sought by, 14–17,
27
objectivity sought in, 30
Science Times, 181
Scientific American, 80, 129
scroll-like structures, 110, *111*

sculpture, 32–33, 158, *169, 177*
Selfish Gene, The (Dawkins), 117
self-organizing chaos, 25–26, *26,*
108–14, *109, 110, 111, 113, 114*
chemical reactions and, *141*
weather and, *56*
self-similarity, *13,* 25–26, *26,* 68,
144–45, 169
in art, 30, *88–89,* 148–49, *168,*
172, 174
of fractal landscapes, 53
in organisms, *70*
in physics, *88*
see also scaling; symmetry
sensitivity of dynamical systems, 18–
19, 21, *90*
avalanches, *47*
feedback and, 19–20, 116
weather, 15–17, 56–60, *56, 57, 58,*
143
Sierpinski arrowhead, 68
Sierpinski curve, 70
Sierpinski gasket, 68
silicone, *14, 97*
Silver, Rollo, *79*
Siva, 16
Siver, Peter, *14*
Skarda, Christine, *128*
Sketch of the Past, A (Woolf), 179
Sligo, Ireland, *113*
smell, sense of, *128,* 171–72
Smith, Homer, *76,* 80–81, *81,* 123
"Snowflake" (Dorn), 93
snowflakes, *14,* 59, 94, *95*
software, for creating fractals, 182–
183
solar system, 47, 49–54, *51, 53,* 54,
138–39
solitons, 112–14, *113*
Sommeria, Joel, *114*
Soviet Ministry of Health, 110
speech patterns, 129
spiral nebulae, *110*
spiral structures:
in chemical reactions, *111, 141*
in ion activity, 145
in organisms, *97, 109,* 110
in Stone Age structures, *113*
Staller, Jane, 80
starfish, *40*
Star Trek II: The Wrath of Khan,
84, *84*
Star Wars, 84, *84*
steam engines, 116

Stewart, Ian, 43
Stone Age structures, *113*
strange attractors, *121,* 137–45, *141,*
142, 143, 144–45, 169, 171–72
in architecture, *171–72*
art and, 166, 172
chemical reactions and, *108*
coinage of term, 139
creating on home computers, 183
of geology, *141*
of human body, 31–32, *31,* 126,
128
Mandelbrot set and, 78
of planetary motion, 139–40
weather and, 59–60, *104–5,* 140
"Strange Attractors: The Spectacle
of Chaos," 166
Sun, 138
suprematist painters, 177
Swift, Jonathan, *41*
Swinney, Harry, *114*
Sydney University, *95*
symmetry:
and chaos hybrids, 93–98, *95, 97*
in organisms, *97*
in physics, *88*
in snowflakes, *95*

television, fractal techniques in, 84
television camera, 116, *117*
Texas, University of, *114*
thermodynamic chaos, 17
Thomas, Lewis, 165
Tiamat, 16
tigers, Bengal, *101*
tornados, 112
toruses, 138–39, *142*
transition areas, 21
trees, *69, 103*
fractal imitations of, 85
trickster characters, 17
turbulence, 112, 113, 131–36, *132,*
133, 135, 136
twins, identical, 19
Typhoon (Conrad), 131

Ueda strange attractor, *142*
Utrecht, University of, *135*

Vague Attractor of Kolmogorov
(VAK), 139
Vak, 139
van der Rohe, Mies, *170*
Van Eenwyk, John R., 180

Van Gogh, Vincent, 166, *169*
Velarde, Manuel, *97*
Vesalius, Andreas, 124
video feedback, 116, *117*
Vinci, Leonardo da, 27–28, *136*
vines, *23*
Visualization Systems Group,
Thomas J. Watson Research
Center, 154
volcanos, *100, 141*
Voyager 1, 53, *53*
Voyager 2, 53

Wallace, Alfred Russel, 108
Washington, *100, 102*
water:
cloud system and, *118*
fractal landscapes and, *102*
fractal motion of, 18–19, 20
soliton waves in, 112–14, *113*
turbulence in, 131, 134, *135*
"Waterfall in Yoshino" (Hokusai),
173
waterspouts, *111*
Thomas J. Watson Research Center,
154
Watt, Thomas, 116
waves, 110, 112–14, *113*
weather, 13–15, *111,* 112
feedback in, 117–18, *118*
long-range forecasting of, 15–17,
56–60, *56, 57, 58, 143*
scaling in, 23–24, 59–60
strange attractors and, 59–60,
104–5, 140
Weierstrass, Karl, 64
Weston, Edward, *105*
Wiin-Nelson, Aksel, 59
wind tunnels, *134*
Winfree, Arthur, *109*
Wisdom, Jack, 49, 52–53
Wolberg, Lewis, 37
woodcuts, *173*
woods, *103*
Woolf, Virginia, 10, 99, 179
worlds within worlds, *see* scaling
Wuorinen, Charles, 180–81

Yellowstone Falls, *102*
yin/yang concept, 16, *142*
Yorke, Jim, 12

Zhabotinskii, Anatol, 110